BAOFENG RADIO THE BIBLE 10 BOOKS IN 1

The Ultimate Guerrilla's Handbook to Master Communication for Emergencies and Outdoor Adventures With Frequencies Hacks, Expert Tips, and Advanced Techniques

By

WARWICK HARDY

© Copyright 2023. All Rights Reserved.

The publication is sold with the idea that the publisher is not required to render accounting, officially permitted or otherwise qualified services. This document is geared towards providing exact and reliable information concerning the topic and issue covered. If advice is necessary, legal or professional, a practiced individual in the profession should be ordered.
- From a Declaration of Principles which was accepted and approved equally by a Committee of the American Bar Association and a Committee of Publishers and Associations.
In no way is it legal to reproduce, duplicate, or transmit any part of this document in either electronic means or printed format. Recording of this publication is strictly prohibited, and any storage of this document is not allowed unless with written permission from the publisher—all rights reserved.
The information provided herein is stated to be truthful and consistent. Any liability, in terms of inattention or otherwise, by any usage or abuse of any policies, processes, or directions contained within is the sole and utter responsibility of the recipient reader. Under no circumstances will any legal responsibility or blame be held against the publisher for any reparation, damages, or monetary loss due to the information herein, either directly or indirectly.
Respective authors own all copyrights not held by the publisher.
The information herein is offered for informational purposes solely and is universal as so. The presentation of the information is without a contract or any guarantee assurance.

The trademarks that are used are without any consent, and the publication of the trademark is without permission or backing by the trademark owner. All trademarks and brands within this book are for clarifying purposes only and are owned by the owners themselves, not affiliated with this document.

Table of Content

BOOK 1: INTRODUCTION TO BAOFENG RADIOS...5
   HISTORY AND EVOLUTION OF BAOFENG RADIOS 5
   TYPES AND MODELS OF BAOFENG RADIOS 5
   WHY USING BAOFENG RADIOS IS A GOOD IDEA..16
   TARGET AUDIENCE AND USER GROUPS 18
   SETTING UP YOUR BAOFENG RADIO 18

BOOK 2: BAOFENG RADIO FEATURES AND FUNCTIONS 22
   UNDERSTANDING RADIO FREQUENCIES AND BANDS..................................23
   CHANNELS, FREQUENCIES, AND CTCSS/DCS CODES 23
   DISPLAY AND KEYPAD FUNCTIONS 24
   AUDIO AND VOLUME CONTROLS 26

BOOK 3: BAOFENG RADIO OPERATION AND PROGRAMMING.....................27
   PROGRAMMING CHANNELS AND MEMORIES 28
   SCANNING AND DUAL WATCH FEATURES 29
   REPEATER USAGE AND OFFSET SETTINGS 31
   EMERGENCY AND NOAA WEATHER ALERTS 32

BOOK 4: BAOFENG RADIO ACCESSORIES 34
   HEADSETS AND MICROPHONES 34
   CARRYING CASES AND HOLSTERS...35
   PROGRAMMING CABLES AND SOFTWARE 36
   UPGRADING AND MODIFYING YOUR BAOFENG RADIO 38

BOOK 5: RADIO ETIQUETTE AND COMMUNICATION PROTOCOLS 41
   EFFECTIVE COMMUNICATION TECHNIQUES 43
   USING RADIOS IN DIFFERENT SCENARIOS (E.G., EMERGENCY, HIKING, EVENTS) 45
   PRIVACY AND SECURITY CONCERNS..48
   TROUBLESHOOTING COMMON COMMUNICATION ISSUES 51

BOOK 6: BAOFENG RADIO MAINTENANCE AND CARE 53
   BATTERY MANAGEMENT AND CHARGING 54

STORING AND PROTECTING YOUR RADIO 58

FIRMWARE UPDATES AND SOFTWARE MANAGEMENT 58

BOOK 7: ADVANCED BAOFENG RADIO FEATURES 62

CROSS-BAND REPEATING 65

REMOTE OPERATION AND PROGRAMMING 73

BOOK 8: BAOFENG RADIO APPLICATIONS 75

COMMERCIAL AND PROFESSIONAL APPLICATIONS 76

PUBLIC SAFETY AND EMERGENCY SERVICES 78

MILITARY AND GOVERNMENT USE 81

RECREATIONAL AND OUTDOOR ACTIVITIES 83

BOOK 9: BAOFENG RADIO LEGAL AND REGULATORY CONSIDERATIONS 86

FCC REGULATIONS (FOR THE UNITED STATES) 87

INTERNATIONAL REGULATIONS AND LICENSING 89

FREQUENCY BAND USAGE GUIDELINES 91

INTERFERENCE AND COMPLIANCE 92

BOOK 10: FUTURE OF BAOFENG RADIOS AND RESOURCES 95

BAOFENG RADIO COMMUNITY AND RESOURCES 97

USER GROUPS, FORUMS, AND ONLINE COMMUNITIES 101

RECOMMENDED READING AND ADDITIONAL REFERENCES 104

CLOSING REMARKS AND FUTURE OUTLOOK 106

CONCLUSION 108

BONUS 110

BONUS 1 110

BONUS 2 110

BONUS 3 110

BONUS 4 111

BONUS 5 111

BOOK 1: INTRODUCTION TO BAOFENG RADIOS

HISTORY AND EVOLUTION OF BAOFENG RADIOS

The Chinese manufacturer BaoFeng made the portable Baofeng UV-5R radio i.e. you can carry it anywhere with your hand. Although it isn't their only model, this was the first globally successful dual-band radio (VHF/UHF) of a Chinese brand. The good news is that it is affordable and easy to use, regardless of whether you are a novice or a pro.

This model came into limelight in 2012 and sold to markets globally. However, the radio lacked FCC Part 95 certification in the United States, therefore, making it unauthorized for use in the GMRS and FRS, only for amateur radio. Note that since 2012, Baofeng has produced many other models based on the UV-5R technology. Other radios from other Chinese producers ensured that their product, e.g the Retevis RT5 shared similarities such as the UV-5R.

Sadly, there have been complaints concerning the UV-5R from several telecommunications regulators. These complaints include issues with frequency interference. This is why countries like Switzerland, Germany, Poland, and South Africa banned it from sale and use. The German Federal Network Agency also rejected the device because it doesn't properly dampen harmonics, disturbing other users. Moreover, the Independent Communications Authority of South Africa issued a ban on the product due to being responsible for radio frequency interference, and regular tuning capabilities; a function that demands am operator Frequency allocation license before buying or using the product.

TYPES AND MODELS OF BAOFENG RADIOS

If you're on a budget, Baofeng radios are your best buddy, especially if you're a ham radio buyer. However, this product isn't limited to one model, as mentioned earlier. Therefore, you might be confused about how to buy the best Baofeng ham radio.

Don't worry, we have sorted that out already. All you have to do is just read the features, pros, and cons of each model so that you can know which one suits your taste. So let's get started and explore these radio models in our magic carpet ride.

1. Baofeng BF-F8HP Dual Band Two-Way Radio

This is a third-generation ham radio with an elevated 8-watt power output, including a high-gain V-85 antenna. Some people may dislike it as a 7" dual-band antenna because it correlates with the required size for UHF (Ultra High Frequency) at 70 centimeters.

For individuals who have been a fan of Baofeng ham radios, using the BF-F8HP is a sign of an upgrade. According to Baofeng, BF-F8HP is a successor of the initial UV-5R model. This makes Baofeng BF-F8HP the most effective ham radio from the company.

While the highest output this model can take is 8 watts, there are two more power settings such as the Low and Medium with 1 watts and 4 watts outputs. As a result, the kind of output you use will change the transmission power.

This model's frequency range starts from 136 - 174 MHz for VHF and 400 - 520 MHz for UHF.

Moreover, for the UHF 70 centimeters band, the recommended frequency range is 430 to 440 MHz. You'll get 400 - 520 MHz for the UHF band from this model. Therefore, don't just be hasty to use it. Ensure you program it first. Furthermore, don't be tempted to exceed a frequency above the required range. You will only end up implicating yourself.

The Baofeng BF-F8HP's transmission mode involves the semi-duplex mode (or simple), enabling you to transmit and receive, but not at the same time.

If you want to enjoy this model's 2000mAh battery for a day, then keep using it at medium power (5 watts), however, you can only enjoy it for 20 hours at a max power of 8 watts.

Pros of Baofeng BF-F8HP

- The coverage range is boosted by significant output power and its substantial gain V-85 antenna.

- The battery life is commendable, so you can use it if you have emergencies or during outdoor activities such as hiking.

- With a third frequency ranging from 65 - 108 MHz, you can get access to a professional FM radio.

- You can decide to select which of the three power levels you prefer. This is another great option to consider when involved in outdoor activities.

Baofeng BF-F8HP drawbacks

- It lacks a programming cable, so you must buy it separately, especially the programming cable from FTDI.

- In the speaker segment, the radio performs less well because of its 700mW sound output. Therefore, I cannot use it for loud zones.

- It doesn't have a larger antenna.

When purchasing this model, there are different variables that come with it for your total satisfaction. They include the following:

- The gadget (radio unit BF-F8HP).
- V-85 antenna with high gain.
- Power adapter.
- 2000 mAh battery.
- Belt clip.
- Earpiece.
- Battery charger.
- Hand strap.
- User manual.

2. BaoFeng UV-82HP High Power Dual Band Two-Way Radio

BF-F8HP also has a relative from Baofeng that has a high power output. This is where UV-82HP comes into the limelight. It also belongs to the class of the 8-watt output of power. Most people are aware that Baofeng ensured that the UV-82 series was for commercial purposes as well as for users of amateur radio.

Nevertheless, the UV-82HP is an upgrade from the original UV-82 series. The difference is that it is specific for only amateur ham fans.

Another difference is its power output of 7/8 watts, which enables good range. You can also be glad that it features the high-gain antenna V-85! As a result, this high power output and an excellent dual-band antenna will give you long-range transmissions!

This radio has three various power settings such as – 1 watts, 4 Watts, and 7/8 Watts. If you feel like using the VHF range, then use the 8 Watts power output, however, regarding the UHF range, consider using 7 Watts. Moreover, this radio also has three various frequency ranges including – UHF (400-520 MHz), VHF (136-174 MHz), and Commercial FM radio reception range of 65-108 MHz.

So, whenever you feel like using the UHF range, configure your handheld radio. via its keypad, a PC, or a laptop to be between the range of 430-440 MHz. However, utilizing a laptop or a desktop requires a cable and programming software.

One amazing feature that comes with this radio is its dual PTT button. It has a switch you can easily flip up (Forwarded to Frequency A) and down (Forwarded to Frequency B) to enjoy two different frequencies. Do you know you can even reset the dual PTT switch to function as a single PTT button?

Additionally, the UV-82HP can also send DTMF tones, enabling you to send caller ID or ANI to commands that require DTMF in most cases.

- It operates in the semi-duplex transmission mode, which lets you send and receive at different times, just like the majority of amateur ham radios.

- Benefits of the Dual Band Radio UV-82HP

- The range is increased by the excellent gain of the V-85 V-85 antenna and the high power output.

- It has 128 channels of programmable memory.

- It contains three various power settings and based on the power output, your battery life can last longer.

- Its dual PTT button enables you to send to two various frequencies.

- It features a 1 Watt speaker output

The Dual Band Radio UV-82HP drawbacks

- This radio includes one 1800 mAh battery, as well as the box lacks any spare battery.
- There is no programming cable, and the keypad is slightly jam-packed.

When you finally buy this model, expect to get the following items:

- The UV-82HP radio.
- One power adapter.
- One battery charger.
- 1 earpiece.
- One hand strap.
- One belt clip.
- 1 user manual.

3. Baofeng 8-Watt UV-5RE+ Ham Radio

The frequency ranges of this model include From 65 to 108 MHz for FM radio, 136 to 174 MHz for VHF, and 400 to 520 MHz for WiFi (UHF). You can use it for a simple or semi-duplex mode of transmission.

Don't just be hasty in using this radio, first program the ham, which can be done via the keypad. Also, it lacks a programming cable, so consider buying it separately. Moreover, to save money, try Getting rid of the cable altogether.

The 2800 mAh battery lets you use it hands-free and has a VOX function. It also has an LED light, a low battery alert, 50 CTCSS codes for security, a keypad lock, and more.

The good things about the Baofeng UV-5RE+ radio

- With the GAMTAAI high gain antenna, you can make it work farther.
- The battery is higher for extended operation hours.
- It includes a hands-free operation.

Bad Things About the Baofeng UV-5RE+ Radio

- It doesn't have squelch technology to clean up the wave frequencies.

- It lacks a programming cable, so consider buying it separately.

- It's slightly difficult to program with the keypad.

Below are the items to expect if intend to purchase this product:

- If you want to call someone, the Baofeng UV-5RE+ radio could be used.

- A GAMTAAI NA-771R Antenna.

- An audio tube earpiece made by GAMTAAI.

- A standard antenna.

- 2800 mAh of power.

- One hand strap and a belt clip.

- An adapter and 1 desktop charger.

A guide for using something in English.

4. The Baofeng UV-5R two-way radio has two bands and a TIDRADIO 771 antenna.

If you're looking for a radio with a voice output that is crystal clear, then this model is for you. The magic behind this amazing feature is thanks to the popular and effective squelch technology that offers quality voice commanding.

This clear sound can be achieved when the wave frequencies are cleaned up by the squelch technology. So, if you find yourself somewhere with bad signals and you're using radio communications, you can still hear what the other person is saying and send your message.

Unlike other models we have discussed earlier, The Baofeng UV-5R Walkie-Talkie features a programming cable so you can program your radio to your preference. Yet, ensure you have a software program on your PC. However, if you lack programming software on your PC, the radio's keypad can come in handy.

You can even use it to reject calls that aren't necessary. If your battery is dead, don't worry, it also has a spare battery and a power-saving mode. Furthermore, the radio has 128 channels and you keep the frequencies you use most.

However, just like the rest we have discussed so far, it also sends data in simple or semi-duplex mode.

Advantages of the Baofeng UV-5R

- Walkie-talkie is its high gain. Radio antenna TIDRADIO 771 boosts the coverage range.
- Improved battery life, and a spare battery for emergencies.
- It features a feature that lets you control how much power it uses and squelch technology for voice compression.
- It has a programming cable so you can save your money for buying a programming cable.
- You can listen to radio frequencies of 65 MHz - 108 MHz.

The Baofeng UV-5R walkie-talkie has the following features:

- When it comes to low power, 4 to 5 watts is very encouraging. More output would have boosted its range greatly. Yet, 4-5 watts is often the required output.
- You can only scan three beats per second.

After buying this radio, expect to see the following item in your package:

- A TIDRADIO 771 antenna with a lot of gain.
- A radio mic.
- Power adapter.
- 2 x 1800 mAh Li-ion batteries that can be charged.
- A Baofeng programming cable.
- Earpiece.
- Belt clip

- Battery charger.

- User manual.

5. The Baofeng Black UV-5R V2+ is a two-way radio with dual bands.

This radio is among the many versions of the UV-5R type. Also, This two-way radio, the Baofeng Black UV-5R V2+ Plus (USA Warranty), is the same as the Baofeng UV-5R Dual Band Two-Way Radio (Black) below (#5).

However, some upgrades have been made to improve it. Let's first talk about the similarities. It shares similar traits, The FM transceiver has 128 channels, priority channel scanning, 104 DCS, and 50 CTCSS codes to keep your communications private, and its frequency range is from 65 MHz to 180 MHz.

Indeed, the features are very similar! Expect to have the VHF range from 136 MHz to 174 MHz and UHF ranges from 400 MHz to 520 MHz. It also has a built-in VOX function for hands-free use.

So what makes the former different from the latter?

The following are their differences:

- It features a better LED screen that contains two colors with a brighter screen.

- It includes a metal casing so that the radio can be heavier.

- The UV-5R comes with 3800mah and 3600mah extended batteries.

There's no point denying that the extended batteries are an advantage for the Baofeng UV-5R Dual Band Two-Way Radio (Black) and the Baofeng UV-5R V2+ Plus (USA Warranty) are both two-way radios.

While some people may not like the metal casing and heavy build, it all depends on how you use a radio.

This radio uses a simple or semi-simple gearbox mode, enabling you to not transmit and receive at the same time. Don't forget to turn on the radio before programming it. You can program the keypad or use a computer and a cable to connect to it. Just as in the previous model, having both a programming cable and a cable doesn't mean you won't require programming software.

The Baofeng Black UV-5R V2+ Plus has a good deal of Dual-Band Two-Way Radio Support to make batteries last longer.

- Hands-free operation is ideal for drivers.
- It's made of metal, so heavy.

Cons of Double-band, two-way Baofeng Black UV-5R V2+ Plus radio with a warranty in the USA

- Bad antenna and output with low power resulting in low range.
- At 700 mW, there is bad speaker output.
- You can only use the added 1800 mAh battery for 12 hours.
- No spare battery.

After buying this radio, expect to see the following items in your package:

- The radio Baofeng UV-5R V2+.
- Another 1800 mAh Li-ion battery that can be charged.
- An antenna.
- An earphone mic/headset.
- An adapter.
- A desktop charger.
- A hand strap and 1 belt clip.

6. The Baofeng UV-5R is a two-way radio with two bands

This radio is affordable, compact, has lots of channel capacity, and is simple to program, which is why most amateur radio fans loved it in the past. While most individuals have upgraded to the 3rd The second generation Baofeng UV-5R model is the same as the first generation Baofeng BF-F8HP model. still famous.

When you get this model, expect to see three various frequency ranges, including:

- From 65 MHz to 180 MHz for FM.

- From 136 MHz to 174 MHz for VHF.

- 400 MHz to 520 MHz for UHF.

You also should avoid using it if you haven't programmed it yet if you don't want to have any issues. This is because the frequency range is beyond that of a regular ham radio. Programming can simply be done with the radio's keypad, PC, or laptop. Don't forget that using a PC/laptop to program this radio requires a cable and software.

If you want your communication to be private then this radio has got you covered with its codes for 50 CTCSS and 104 DCS, which are security codes.

Additionally, you can Use your hands to scan priority channels and operate through voice-operated transmitting as expected, it features a simple or semi-duplex mode of transmission to send and receive at different times. Simultaneous communication is not possible.

Pros of the Baofeng UV-5R Two-Way Radio Double Band

- The radio features a built-in VOX function for hands-free operation. VOX means Voice Operated Exchange.

- It includes a built-in flashlight.

- You'll get emergency and low-battery notifications.

- It includes several power outputs with a maximum of 5 Watts and a minimum of 1 Watt and 4 Watts.

- It isn't heavy and simple to use.

Cons of Baofeng UV-5R Dual Band Two-Way Radio

- At 700 mW, expect a bad speaker output which is not ideal in noisy areas.

- You only have 12 hours to use it and there's no spare battery. So you cannot rely on it in emergency situations. The only solution is to buy another battery in case the built-in one fails you.

- The highest this radio power output can take is just 5 watts and it lacks a high gain antenna. Even the antenna is not encouraging!

After buying this radio, expect to see the following items in your package:

- The Baofeng UV-5R Dual Band Two-Way Radio.
- One 1800 mAh Li-ion battery.
- One antenna.
- One headset/earphone mic.
- One adapter.
- One desktop charger.
- One belt clip and one hand strap.

Also, note that there's an English manual that comes with this radio.

7. Two-Way Radio Dual-Band Transceiver Baofeng UV-5RA Ham

If you're a novice you should try this small handheld ham radio. However, don't expect to get excellent coverage, cool features, etc.

Don't be sad because it's still a member of the UV-5R series family, so you can still enjoy some features. For example, 400 to 136 MHz for VHF 136 to 174 MHz for UHF, and 52 MHz for UHF are available.

The radio can handle 128 channels and can be used hands-free with the VOX function, as well as 50 DCS codes and 104 CTCSS codes to secure your communications.

If you like, consider tuning between 65 and 108 MHz, which is a commercial FM radio.

The good news is that it features a double PTT (push-to-talk) button, for transmitting on two different bands and programming the radio. Nevertheless, it's better to use a computer and programming cable if you want to program the radio. Just endeavor to purchase the cable separately.

The highest power output this radio can take is 4 Watts and a less powerful output of 1 Watt. When using it, expect the battery to last from 9 to 12 hours.

Just like others, It can also work in a simple or semi-duplex transmission mode. for transmitting and receiving, however, not simultaneously.

Pros Baofeng UV-5RA Two-Way Radio

- With two PTT keys, it's easy to use.
- It features a VOX function for hands-free operation.
- It's very affordable.

Problems with the Baofeng UV-5RA two-way radio

- Bad antenna and power output, resulting in poor coverage.
- Bad build quality. As a result, it won't go well for one sharp fall.
- Discouraging battery.

After buying this radio, expect to see the following items in your package:

- One UV-5RA Baofeng Radio that works both ways.
- It has one 1800 mAh battery.
- One belt clip and one hand strap.
- One adapter and one charger for your desk.
- One earphone.
- One antenna.
- One manual in the English language.

WHY USING BAOFENG RADIOS IS A GOOD IDEA

Baofeng two-way radio is still the major and most effective type of communication for many workers and individuals worldwide. Regardless of whether you work as a police or in a factory, baofeng two-way radios, you can heavily depend on it. Moreover, it's durable so you can enjoy it for a very long time. You'll learn many reasons to use baofeng radios in your daily activities.

No maintenance and service fee

If you're worried that after buying this radio, you ought to maintain and service it which can be expensive in the long run, relax, because baofeng radios have no added fees. So you can now save the money you might be spending on maintenance and service fees. This is one of the reasons it's better than cell phones that will change you for using its service. This radio is a cheerful giver, so enjoy it.

Improved worker safety

If you want your workers to feel safe due to the workplace environment, then these two-way days can come in handy, especially during emergencies. Even the VDC made a study and discovered that approximately 65% of respondents testified of the protective abilities of this radio on workers' safety. It was during an emergency to quickly profer a solution to workers before it escalated.

Reliable battery life

While most cellphones are improving in their battery performance, there is still limitation during peak usage. As a worker, using your cellphone to communicate while working might be disappointing due to low battery life. You will realize that you cannot rely on the battery for a full shift.

However, baofeng radio batteries are stronger than cell phone batteries. It can last for more than 12 hours in peak usage. So if you are working drying your shift, you have no worries of interrupted workflow. As a result, there will be ease of communication and improved performance in the workplace.

Baofeng radios are strong and reliable, and they are also flexible and portable, which makes them less likely to get broken. Because of this, you can listen to this radio for a long time. Furthermore, you won't have to pay to repair it since it's flexible and durable. VDC made a study and discovered that approximately 18–20% of cellphones may not function again within a year, meanwhile only 4–8% of two-way radios might also not function. As a result of relying on cell phones instead of two-way radios, it reduces performance and results in losses.

Audio quality

Most factories and industries suffer from noise pollution, which hinders effective communication among individuals. Nevertheless, clear communication is still necessary and cell phones cannot save the day in such situations. This is one of the reasons baofeng radios were made with an enhanced voice quality by tested software algorithms to reduce noise. As a result, this radio should be encouraged in factories and industries.

Group communications

In the case where employee collaboration is needed, the baofeng radios have a 'one too many' feature that allows for group communication and the ability to connect many people.

Features and functions that add value

Modern two-way radio digital radios such as the baofeng two-way radio offer new high-end features like messaging, FM radio, and GPS. In addition to these features, other workplace applications, which include location tracking are an added value to the radio.

TARGET AUDIENCE AND USER GROUPS

When looking for two-way radio communications for personal and professional usage, think of Baofeng radios. It is generally used because of its affordability and its ability to function well in diverse user groups and target audiences. Amateur radio operators, emergency contact fans, outdoor adventurers, hikers, and campers are the main groups of people who buy this product. Baofeng radios are also often used in security, hospitality, and event management, among other fields.

SETTING UP YOUR BAOFENG RADIO

You can set up your Baofeng UV-5R by hand from the radio's keypad by following the steps below.

How to set up a simplex line by hand

Step 1: Press [VFO/MR] to switch to tone mode.

Step 2. It is essential to press [A/B] when trying to choose the A-Side (upper display).

For the radio to work, the A side ought to be used for the stations set up. The programming information that you enter on the (lower display) that is B side will not be saved.

Step 3: To change the frequency band, press [BAND].

If you intend to select either 136 MHz (VHF) or 470 MHz (UHF), flip the [BAND] switch.

The radio will stop working if the wrong band is picked in the 5th Step.

4. Turn off TDR (Dual Watch/Dual Standby).

It is vital to press [MENU] 7 that is [MENU] [press the up and down Keys for arrows OFF [MENU] [EXIT]

When you are programming from the radio, it is strongly advised that you turn off TDR.

Step 5: Type in the number.

You can tell the radio what channel to use by pressing the keys.

Step 6: You can enter the send CTCSS/DCS code.

13 on the CTCSS menu [MENU] [Type in or Pick Up Code XXXX] [MENU] [LEAVE] DCS 12 [MENU] [pick code XXXXX] [MENU] [LEAVE]

Step 7: Put the frequency to a channel.

[MENU] 27 Type in "channel number XXX" from the menu. [MENU] [LEAVE]

How to set up a rebroadcast channel by hand

Step 1: To switch to tone mode it is essential you press [VFO/MR].

Step 2: To pick the Side A (top screen)Press [A/B].

Just like with the simplex channels, you need to use the A side to set up the rebroadcast channels on the radio. The programming information you put on the lower display will not be saved.

Step 3: To change the frequency band, press [BAND].

To pick either 136 MHz (VHF) or 470 MHz (UHF), flip the [BAND] switch.

The radio will stop working if the wrong band is picked.

Step 4: If you want to, you can clear CTCSS/DCS codes that were previously given.

If you don't have any old codes or are starting the channel and don't need any codes, turn off the following menu items.

- [MENU] RX DCS 10 [MENU] [type 0 to turn it off] [MENU] [EXIT]

- [MENU] RX CTCSS 11 [MENU] [type 0 to turn it off] [MENU] [EXIT]

[MENU] TX DCS 12 [MENU] [type 0 to turn it off] [MENU] [EXIT]

TX CTCSS - [Main Menu] 13 [MENU] [type 0 to turn it off] [MENU] [LEAVE]

Step 5: Turn off TDR (DualWatch/Dual Standby).

Click on it [MENU] 7 [press the up and down arrow keys] OFF [MENU] [LEAVE]

When programming from the radio, it is strongly suggested that TDR be turned off.

Step 6: Delete any data that is already on the channel.

If this is your first time setting up the channel, skip this step.

Press [MENU] 28 and then use the up and down arrow keys to pick a station. [MENU] [LEAVE]

When programming from the radio, it is strongly suggested that TDR be turned off.

Step 7: Type in the frequency of the repeater's signal for usage.

You can tell the radio what channel to use by pressing the keys.

Step 8: Enter the frequency shift for the repeater.

[MENU] 26 [enter the offset for a 70 cm or 2 m repeater] [MENU] [LEAVE]

Step 9: Type in the changes for the transmit frequency.

Click on it [MENU] 25 [MENU] [type 1 for upside-down shift or 2 for upward shift] [EXIT] [MENU]

Step 10: You can enter the send CTCSS/DCS code.

The CTCSS menu 13 [MENU] [Type in or Select Code XXXX] [MENU] [LEAVE]

[MENU] DCS 12 [MENU] [pick code XXXXX] [MENU] [LEAVE]

Step 11: Give the channel the receive frequency that you set in Step 7.

[MENU] 27 [Type in channel digits XXX] [MENU] [LEAVE]

Step 12: Press and hold the [Scan] button to turn on Reverse Mode and see the broadcast frequency.

Step 13: Give the channel the broadcast frequency.

Click on it 27. [MENU] [Insert the exact memory channel you used in step 12] [MENU] [LEAVE]

Step 14: To leave Reverse Mode, press the [Scan] button.

Just do the steps above again to add more channels. If you follow these steps carefully, you should be able to set up all 128 channels (000–127) on your Baofeng UV-5R the way you want.

BOOK 2: BAOFENG RADIO FEATURES AND FUNCTIONS

OVERVIEW OF RADIO COMPONENTS

Many people like how the Baofeng UV-5R radio system works because it can be used in many situations and has many useful features. In this case, it is popular for amateur radio fans, emergency contact workers, and people who like to go on outdoor adventures. The design of this radio includes a number of important parts that make it work and be useful.

1. Antenna

To begin, the antenna is a very important part of both sending and receiving radio messages. It makes sure that communication works well over long distances and rough terrain, so users can stay linked in a variety of settings. The UV-5R's receiver is designed so that it works reliably even in tough situations.

2. Screen for show

The second thing about the radio is its clear and useful display screen. The fact that this screen displays vital information such as channel frequencies, battery life, and current settings makes it a crucial component of the interface. The display improves the experience by making the content simple to read and providing instant access to crucial data.

3. Keypad

The keypad is another important part that lets users quickly enter orders and change settings. Navigation through the radio's different functions is easy thanks to its simple design. Users can easily change bands, turn on scanning modes, and change settings to suit their needs.

4. A battery

The UV-5R can be used for extended periods of time during emergencies because its battery can be removed and charged. When you utilize this capability, users will be able to carry additional batteries with them so that they can continue talking in the absence of a power source. The radio's power management system also helps in making the best usage of the battery, and this increases the device's total runtime.

5. A speaker and a mic

Furthermore, the speaker and microphone parts are necessary for clear conversation. A good speaker makes sure that the sound is clear, so users can hear texts and information clearly even when there is noise around. The microphone makes it easy for voice signals to be sent smoothly, so users can send notes that are clear and effective.

6. Ports

The UV-5R is very flexible because it has many ports, such as charging ports, recording jacks, and programming ports. These ports make it easy for users to charge the device, connect external audio devices for better sound output, and program the radio to meet specific conversation needs. This makes the radio more useful and flexible for a range of situations.

UNDERSTANDING RADIO FREQUENCIES AND BANDS

People talk with each other differently while utilizing the Baofeng Radio UV-5R because it works on two bands which include UHF and VHF. The VHF band is between 136 and 174 MHz, and the UHF (Ultra High Frequency) band has frequencies between 400 and 520 MHz. Users are able to connect to each other in various ways using frequency bands, it contains about 400 to 520 MHz UHF, and in terms of VHF the band is about 136 to 174 MHz.

The UV-5R can reach a lot of channels because it has wide frequency coverage. This makes it good for different communication needs, like amateur radio, emergency services, outdoor activities, and professional use. The UV-5R can be programmed to reach certain frequencies within these bands, which makes communication possible in a variety of settings.

CHANNELS, FREQUENCIES, AND CTCSS/DCS CODES

A well-known dual-band handheld transmitter is the Baofeng UV-5R. It works with frequencies between 136 and 174 MHz and 400 and 520 MHz. As many as 128 channels can be stored, and it works with both CTCSS (Continuous Tone-Coded Squelch System) and DCS (Digital-Coded Squelch) codes for selective squelch. These bands and CTCSS/DCS codes are used a lot:

The frequencies:

The VHF range is from 136 to 174 MHz.

400 to 520 MHz is UHF (Ultra High Frequency).

Continuous Tone-Coded Squelch System, or CTCSS Codes are used for tone delay that can't be heard so that multiple people can use the same channel without bothering each other. Codes like 67.0, 77.0, 88.5, 100.0, 123.0, and more are used a lot.

They are called DCS codes and are used in digital squelch devices to stop other users from interfering. There are numbers such as 023, 025, 026, 043, 047, and more. It is essential to check the rules and regulations that are set for your local communication body and the codes and frequencies you ought to use depend on it. Always make sure you are following the rules set by your country's radio contact laws.

DISPLAY AND KEYPAD FUNCTIONS

The Baofeng UV-5R's screen and keypad can be used in different ways to reach different features and settings. Here are some of the most popular functions:

How to Use the Display:

- Channel Display: This shows the current frequency of the working channel.

- The signal Indicator Strength: Users can detect how high the level of signal that is being received.

- The Battery Indicator Level: Users can utilize this indicator to know the level of power that is in the battery.

- Menu Navigation: Shows the menu choices that let you get to different functions and settings.

How the Keypad Works

PTT stands for "Push-to-Talk."

- Hold down [PTT] to send, and let go to receive.

[CALL] Key 1 on the side

- Once the user has decided to switch on the FM radio, it is essential to click on the [CALL] button. To switch it off, click on it again. If the user goes through the stations in the FM, press the [SCAN] button. The alarm can also be put off if you press and hold the [CALL] button. If you repeat the process by holding it again, the alarm will go off.

Side Key 2:

- If you intend to switch on the lights, click on the [MONI] button. Press it again if you intend to off it.

- To watch a signal or open mute, press the [MONI] button.

- Button for [VFO/MR]

- To change between these two modes, press the [VFO/MR] button.

- VFO stands for frequency mode and MR for memory recall mode.

[A/B] Push Button

- Press the [A/B] button to change the showing of the frequency.

- The first display, A, shows the FREQUENCY/

CHANNEL/NAME Display

- B Display = lowest TIMER/CHANNEL/NAME Show off

[BAND] Key

- To switch bands in frequency mode, press the [BAND] button.

- Hold down the [BAND] button while the FM radio is on to change the band (65-75MHz or 76-108MHz).

Key [SCAN]

- To use the Reverse feature, press the [SCAN] key. The TX and RX bands will be switched.

- To scan (freq/chan), hold down the [SCAN] key for two seconds.

- Press the [SCAN] key while the FM radio is on to look for FM stations.

- Press the [SCAN] key to look over the RX CTCSS/DCS while you're setting it.

[#] Key

- To change between High and Low power, press the [#] key.

- To lock or open the keypad, press and hold the [#] key for two seconds.

AUDIO AND VOLUME CONTROLS

The UV-5R Baofeng is a well-known handheld transceiver that works on both bands and has many useful features, such as audio and volume settings. If you intend to make changes to the volume as a user, check at the top of the radio you will see a volume knob. If you intend to change from a certain audio station to another, you will see a button at the front of the radio

On the Baofeng UV-5R, these are some of the most important video and sound controls:

1. Volume knob: This is on top of the device and lets you change the sound coming out.

You can change between different stations or frequencies with the channel selector.

3. The squelch setting helps cut down on background sound when the radio is not picking up a signal.

4. Speaker/mic jack: This lets you add an external speaker or microphone to improve the sound quality.

These buttons should make it easier for you to control the Baofeng UV-5R's sound and brightness.

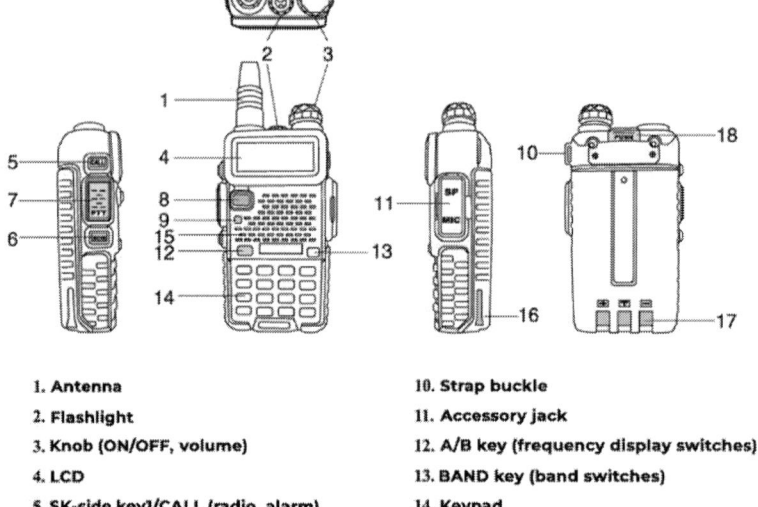

1. Antenna
2. Flashlight
3. Knob (ON/OFF, volume)
4. LCD
5. SK-side key1/CALL (radio, alarm)
6. SK-side key2/MONI (flashlight, monitor)
7. PTT key (push-to-talk)
8. VFO/MR (frequency mode/channel mode)
9. LED indicator
10. Strap buckle
11. Accessory jack
12. A/B key (frequency display switches)
13. BAND key (band switches)
14. Keypad
15. SP. & MIC
16. Battery pack
17. Battery contacts
18. Battery remove button

BOOK 3: BAOFENG RADIO OPERATION AND PROGRAMMING

BASIC OPERATION AND COMMUNICATION

The affordable and useful Baofeng UV-5R is one example of a dual-band handheld radio. Its mechanism of operation and communication should therefore be understood for its maximization.

1. To switch ON hold down the power button till the smartphone starts lighting up Press and hold the power button briefly a second time to switch it off.

2. Setting the frequency: Two choices are available to input the frequency which may involve keying into a keyboard or selecting using a channel selector. Save channels that you commonly use so as to ease your access.

3. Volume and Squelch Control: Rotate the volume knob for level adjustment. While not in use, the squelch level can reduce the background noise.

4. Choose the Channel: With a channel selector, you can select the chat channel. The UV-5R works with numerous bands including several channels.

5. Sending and receiving: To send the message, just push the PTT. Once you release the button, you will listen for messages. Wait briefly before talking if you do not want the start of your speech to be cut short by pressing the PTT button.

6. Changing Settings: Using the menu button and navigation keys, you may change several options such as volume, search parameter, squelch settings, and other mutually modifiable parameters.

7. Sending and receiving signals: Check on signal strength in order to confirm your ability to communicate by looking at it closely. If the signal is weak, you might want to adjust the antenna or move to a better spot for clear communication.

8. Radio Etiquette: Learn the basics of radio contact etiquette, like sending short and clear messages, waiting for a pause before sending, and not disturbing others or sending long messages.

9. Battery Management: Check the battery level often to make sure you don't lose power when you need to be communicating urgently. For long use in the field, bring extra batteries or a portable charger.

10. Legal Considerations: Before using the Baofeng UV-5R, make sure you know the rules and licenses that apply in your area. Make sure you follow the laws and permissions that apply to radio contact in your area.

In the Bonus section, you can find a video tutorial for the main operation and programming procedure.

PROGRAMMING CHANNELS AND MEMORIES

The programming of memories and channels is crucial for users who wish to improve connectivity and specifically target certain frequencies. Users with programming backgrounds may be able to easily save or retrieve a number of customized to their specialized communication channels.

To get their data into the memories or channels, users must follow certain steps to program the Baofeng UV-5 R's frequencies, give channels, and make the settings work best. Programming is easy and after understanding it, you will be able to add the necessary frequencies to the radio's memory. It creates easy, rapid, open, and efficient communication in several operating situations.

Here are the general steps you need to take to set up files and channels on the Baofeng UV-5R:

1. Switch to frequency mode: Once you press the VFO/MR key, you will be operating in frequency mode.

2. Select the A-Side: Press the A/B button to decide on the A-side for programs. The contents of the data in the B Side shall not be stored.

3. Select the band of frequencies, and click on the BAND button for switching between VHF (136 MHz) and UHF (470 MHz) bands.

4. Enter Frequency: Just enter the required bandwidth using the keyboard. Do not forget about the local regulations and band plan.

5. Not required: Enter the DCS or CTCSS code for the desired channel through the appropriate menu items. This can ensure privacy in talking and keep away from other intruders in the conversation.

6. Give each channel a frequency: The menu enables you to connect the specific frequency with your preferred channel number. Such a step ensures that they can obtain the specified frequency, once the device has been initiated.

7. Save the Channel Information: In order to save the programmed channel information to the radio's memory follow the instructions in the user manual.

8. Test the Channel: Checks should be made to ensure the accuracy of the content. Verify that you are able to transmit and receive signals correctly at the selected frequency.

If you need to save more than one, you will repeat those same steps for all the channels in the radio's memory.

10. Sort Memory Channels: The menu function of the radio allows you to arrange memory channels so that they are sortable for easy access during listening.

11. Label the Channels: Give all of the presets names that are easy to remember so that you can quickly find what you need.

12. Save the Configuration: Once your channels are programmed and arranged, be sure to save the setting so that the selected channels remain when you shut down the radio next time.

SCANNING AND DUAL WATCH FEATURES

It is a popular two-band handheld transceiver with advanced scanning and dual watch features that one can use to monitor several channels simultaneously. This is a really useful phone because its flexibility makes it very popular with lots of customers for various contact needs. We will talk about the Baofeng UV-5R's tracking and dual watch features in this part. We will explain how these features work and why users will benefit from them.

1. Radio Frequency Scanning:

With its advanced frequency scanning, the UV-5R enables users to examine a variety of frequencies from the VHF and UHF bands. This feature helps users search for live channels and update themselves about the current happening in their region.

2. Channel Scanning:

Users also get a feature to change quickly among the saved channels pre-stored as an inventory in its memory. The ease with which one can shift between channels in this regard becomes important in keeping tabs on specific frequencies as a crucial aspect of communication management.

3. Dual Watch:

This model of radio provides people with a feature called dual watching, where they can monitor two channels simultaneously. It's a good thing because sometimes you want to hear two channels simultaneously. This will help ensure you do not overlook any important transmissions while monitoring other frequencies for additional contacts.

4. Dual Standby Capability:

Users can monitor two different channels or bands on a single unit by a function called "dual standby" with the UV-5R. In addition, this ensures ease of switching channels. It enhances efficiency in conversation because it is possible for people to update about various communication lines with no negative effects on monitoring precision and reliability.

5. Customizable Scanning Modes:

The UV-5R comes with several configurable scan modes whereby the scanning process suits your specific needs. Users are free to alter the scanning options to match their preferences such as manual or automatic scanning. It simplifies this as the person carrying out the scanning moves faster and is able to complete it in real-time.

6. Adjustable Scan Speed:

For adjustability, the UV-5R comes with a scanning speed setting wherein people can opt to select the speeds at which they run through frequencies or channels. This enables adjustability of the scanning speed in order to ensure that all communications activities that fall within a particular frequency band are comprehensively scanned.

7. Priority Channel Scanning:

This capability allows some channels more urgency for watching as well as fast access to critical frequencies of important contacts. This feature is helpful when some channels should be attended to continuously and responded to quickly.

8. Seamless Integration with CTCSS and DCS:

UV-5R does not have any issues at all when using CTT and DCS signals. This provides private contact and limits interruptions due to transmissions that are irrelevant. As an extra functionality, the scanning and dual watch aspects are further enhanced for smooth and safe communication on multiple networks.

REPEATER USAGE AND OFFSET SETTINGS

Using repeaters and setting the distance are important parts of using the Baofeng UV-5R, especially when talking through repeater systems. For correct use, follow the steps below:

1. To get to Repeater Mode, press the VFO/MR button. This will take you to Frequency Mode.

2. Pick the A-side: Press the A/B button to pick the A Side for programs.

3. Choose Repeater Shift. Ensure that you configure your repeater system correctly using the menu. Set both your transmitting and receiving frequencies properly so as to compensate for the repeater offset.

4. Input Repeater Frequency Pair: Use appropriate frequency offsets while setting the input and output frequencies of the repeater.

5. Check the Offset way: Make sure the offset is going in the right way (positive or negative) according to the repeater's instructions.

6. Test Communication: Send a code through the echo system to make sure the talk works. and make sure it gets sent back.

For using the Baofeng UV-5R with repeater systems to work, it is very important that you fully understand and follow these directions. For exact offset and frequency information, make sure to look at the repeater's manual or the area repeater directory.

How to Set the Plus/Minus (+/-) Offset Duplex for the Baofeng UV-5R Repeater

Note: If you wait more than 9 seconds between any of the steps below, you'll be taken back to the home screen and have to start all over again. So get those fingers moving!

1. Press the Menu key.

2. Type 25 into the keyboard. This will take you to a menu to change whether the duplex is + or – (600 is the default).

3. Again, click on the menu button. "To edit the part of the menu" If you can use the arrow keys to see more choices, a little arrow will show up. You can switch between +, –, and blank in this case, so a choice will show up.

4. Use the up and down arrows to move to the setting.

5. Once more, press Menu. This means "you can now save the changes made in step 4." If a speech recorder is available then you will listen to " confirm" which means success.

6. If "exit" is pressed, one goes back to the home screen; otherwise, after 9 seconds the system automatically loads up the homepage.

Done! You're ready to go with your offset twin.

EMERGENCY AND NOAA WEATHER ALERTS

The Baofeng UV-5R is known for being able to communicate through amateur radio, but it doesn't have built-in support for emergency alerts or direct NOAA weather reports. Users can still get important weather information and emergency alerts through other channels, though, so they can stay informed and ready for anything that might happen.

Understanding Emergency Alert Features:

The Baofeng UV-5R doesn't have any built-in emergency alert features, but users can directly tune in to emergency broadcast channels to stay up to date on local emergencies, disasters, and other serious situations. Users can make sure they stay safe and healthy in tough situations by directly tuning in to emergency frequencies and listening for important updates and announcements.

Going Over NOAA Weather Alerts:

The Baofeng UV-5R doesn't directly support NOAA weather alerts, but users can manually tune in to NOAA weather radio stations to get weather updates and alerts. For those interested in getting up to date with the current meteorological information containing severe weather alerts among others, users need to tune in to the appropriate NOAA weather radio channels. It ensures that they get information on local and weather changes hence prepared.

Other Ways to Get Weather Alerts:

Users can add to the Baofeng UV-5 R's features by using external weather alert devices or smartphone apps that send out emergency and real-time weather reports. Together, these add-ons and apps can be useful extras. They can send timely weather and emergency alerts straight to the user's device, making them more aware of their surroundings and better prepared for bad weather and emergencies.

Adding External Weather Alert Systems:

Users can add external weather alert systems to Baofeng UV-5R to improve their ability to get important weather updates and emergency alerts. By connecting weather alert systems that work with each other, users can make sure that real-time weather information and emergency alerts flow smoothly. This lets them stay aware and take the steps they need to stay safe and healthy.

How to Stay Informed and Ready:

Even though the Baofeng UV-5R might not have built-in NOAA weather alert or emergency alert features, users can still use the device's communication features and connect external solutions to stay aware and ready for emergencies and bad weather. By using both manual tuning and external devices and apps, users can improve their general safety and communication, making sure they stay alert and ready for a wide range of environmental and emergency situations.

BOOK 4: BAOFENG RADIO ACCESSORIES

ANTENNAS, BATTERIES, AND CHARGERS

Many antennas, batteries, and chargers are commonly included with the Baofeng UV-5R, giving consumers ease and versatility in a space of operating situations. An outline of these elements is provided below:

1. Antennas:

A stock, general-purpose antenna is frequently included with the Baofeng UV-5R. In order to improve signal reception and transmission over longer distances or in particular frequency ranges, users may also choose to install aftermarket antennas. These antennas can be made to fit a range of communication needs and tastes by changing in length and design.

2. Batteries:

Rechargeable lithium-ion batteries are normally included with Baofeng UV-5R units. These batteries may be recharged with suitable chargers and are made to have a longer operating life. To guarantee uninterrupted connection over extended periods of use or in scenarios where power supplies are scarce, some users can decide to buy extra batteries.

3. Chargers:

A standard charger is typically included with the Baofeng UV-5R, enabling users to recharge the device's batteries. This charger comes with an option for both AC and DC power sources which makes it more useful for users since they can charge the battery under different circumstances. Additionally, for instance, to quickly recharge multiple devices simultaneously during travel or at home, one may use a portable car charger or multiple units.

HEADSETS AND MICROPHONES

Users can tailor their communication experience to their unique needs and tastes by using the Baofeng UV-5R with a variety of headsets and microphones. The following is a list of microphones and headsets that work with the Baofeng UV-5R:

1. Headsets: A variety of compatible headsets, such as earpieces, over-the-head headphones, and in-ear monitors, are available for users of the Baofeng UV-5R. These are headsets that enable users to communicate with other people without actually handling the device, thus making their interaction very comfortable for them in cases when they cannot do it or its impossible. The users should be able to identify the correct style of headsets for communicating with their comfort.

2. Microphones: There are several microphone options that the Baofeng UV-5R is compatible with, including speaker, shoulder, and portable microphones. These microphones provide improved voice quality and transmission capabilities, enabling users to have productive conversations in difficult or noisy settings. Users can select the kind of microphone that best fits their tastes and communication requirements, guaranteeing dependable and clear transmission during crucial activities.

CARRYING CASES AND HOLSTERS

Carrying cases and holsters are essential for protecting the Baofeng UV-5R, making it easy to tote while engaging in a variety of activities. These add-ons provide consumers with useful ways to carry their gadgets safely because they are made to fit the radio and all of its attachments. The features and advantages of carrying holsters and cases for the Baofeng UV-5R are examined in detail below:

Maintaining Cases:

The Baofeng UV-5R carrying cases are made of sturdy materials that offer dependable defense against impact, moisture, and other external elements. Users may effectively arrange their gadgets and accessories with these cases' customizable compartments and adjustable dividers. Furthermore, some carrying bags have shoulder straps or strengthened handles that allow users to easily transfer their equipment while traveling or conducting fieldwork. Even in harsh conditions, the Baofeng UV-5R is kept safe and well-protected thanks to the sturdy design of our cases.

Holsters:

Holsters provide users with a convenient way to carry the Baofeng UV-5R by allowing them to firmly fasten the gadget to their belts or clothes. These holsters are made to give users quick and simple reach to the radio, so they can quickly get it back in an emergency communication situation. A lot of holsters have strong materials and long-lasting clips, which make it possible to connect and hold the Baofeng UV-5R securely. Additionally, some models have extra pockets or compartments to hold extra batteries, antennae, or other necessary supplies. The Baofeng UV-5R is a great option for consumers who want mobility and simplicity of use in a variety of operating settings since these holsters allow you to utilize your hands-free while still maintaining convenience and accessibility.

Selecting the Appropriate Carrying Option:

The choice of carrying case or holster for the Baofeng UV-5R is contingent upon personal inclinations and particular operational needs. Considerations including robustness, comfort, and ease of usage are important for choosing the best carrying option for each user. When selecting a carrying case or holster, it is critical to take into account the necessary level of protection, the convenience of transit, and the requirement for quick access to the device. Users can choose an accessory of best fits their communication needs and preferences by assessing the design, construction, and extra functions of these devices.

PROGRAMMING CABLES AND SOFTWARE

Among non-professional radio operators worldwide, Baofeng handheld two-way radios are widely recognized. With good cause, models like the UV-3R, UV-B6, BF-F9, and the venerable UV-5R are common fixtures in the contemporary ham shack. These radios are incredibly affordable, small, feature-rich, and simple to operate. They are the least expensive amateur radios available. Because of these qualities, Baofeng is well-known throughout the ham community.

Even for the highly seasoned ham who is unfamiliar with Chinese-made radios, Baofeng radio programming from a computer is not that simple and can get messy. Baofeng offers a USB cable driver that is glitchy, and the manufacturer's programming software is essentially garbage.

Thankfully, alternative solutions exist and can be excellent when put correctly. The only thing you should know is to fix these settings, and at what place you have to search for them. Following this, programming usually becomes quite straightforward.

The instructions below will help you locate and correctly install the programming software and driver for the Baofeng USB Programming Cable. NOTE: Keep following all the steps until the end." Essay One: Hypothesis. Failure to follow through may lead to more frustration if things don't happen as you intended them to.

1. **First things first: computer on, radio off.**

2. **Look for the appropriate USB programming driver based on your OS.**

The Prolific driver for Windows, Mac, and Linux may be downloaded from http://www.miklor.com/COM/UV_Drivers.php. The Windows driver for the Baofeng-UV Series USB Programme Cable can be found in the USB Cable Driver folder on the CD that came with the cable if you bought it from Buy-Two-Way Radios. Navigate to USB Cable Driver > USB Cable Driver Prolific 3.2.0.0.exe after inserting the CD. The right driver is this one.

3. Set up the driver for the USB programming cable.

The USB cord should not be connected at this time.

4. Locate the CHIRP Programming Software for your operating system.

You can download the most recent stable build for Linux, Mac OS X, Windows XP, 7, and 8 from http://chirp.danplanet.com/projects/chirp/wiki/Download. It's also in the CHIRP Programming Software folder on the Buy Two Way Radios CD. Choose the folder that suits your OS. The installation of the software should be done first before launching it.

5. **Directly connect the programmer to the USB port of a computer.**

Slotting the other side of the plug in a socket turns on the radio.

Extra actions for Windows

- After connecting, a notification might or might not appear. Look through the Windows Device Manager to confirm the connection. The Windows version you are using will determine how to access the Device Manager. Click on Ports (COM and LPT) once you're in Device Manager. The Prolific USB-To-Serial Com Port (COMX) should be the one indicated. X is the number of the COM port. This number will be needed when configuring the programming software, therefore, make a note of it and write it down where necessary.

- Should there be a question mark inside a yellow triangle beside a USB device then Windows failed to install the driver correctly. To fix the problem, right-click the device and choose Update Driver Software from the menu. Otherwise, choose the Browse My Computer option for manual installation of the driver.

- Select Permit me to select a device driver from my computer's list. Click NEXT after selecting Prolific USB-To-Serial Com Port Version: 3.2.0.0. The Device Manager's yellow alert ought to vanish and the proper driver ought to be installed. Make a note of the COM Port.

6. Launch your programming software for CHIRP.

Select Download From Radio by clicking Go to Radio in the navigation bar. Select the vendor (Baofeng), the radio model (UV-5R), and the COM Port number for the USB cord using the drop-down menus in the pop-up window. Now that your radio and computer are linked, making interaction between CHIRP possible.

Extra Guidelines for Windows laptops

The device manager on some laptops might show that the cable is installed successfully, but CHIRP might not be able to identify the COM Port. If the device has Windows Power Management turned on, this could happen. Go to Device Manager and perform right-click on the Prolific USB-to-Serial Comm Port (COMX) to enable the cable. X is the number of the COM port. Click Properties from the menu. Select the tab for Power Management. Click Apply or Save after removing the check from the box next to Allow the computer to turn off this device to save power.

Use CHIRP programming software to confirm the connection. In the navigation bar, select Download From Radio. Select the vendor (Baofeng), the radio model (UV-5R, for example), and the COM Port number for the USB cable using the drop-down lists in the pop-up box. The radio frequencies should be downloaded by the software. The frequency table will show up on the screen after the data transfer is finished. You can now use your computer to program the radio.

In order for you to enjoy your radios for many years to come, as an authorized Baofeng Dealer, we want the installation to be as simple and painless as possible. The Baofeng radio should be easy to connect to your PC if these instructions are followed correctly. After the first setup, all you need to do to connect to the computer going forward is plug in the USB programming cable to both your computer and radio, then launch CHIRP.

If you follow the installation instructions but are still having issues, there is now an alternative remedy accessible. Windows XP, Windows 7, and Windows 8 computers can quickly and simply install the XLT Painless Programming Cable. Many systems include automatic installation that requires little or no user participation. With Baofeng, Kenwood, Wouxun, and other two-way radios that have standard two-pin Kenwood connectors, the XLT Painless Programming Cable is compatible.

UPGRADING AND MODIFYING YOUR BAOFENG RADIO

Amateur radio operators and other hobbyists love Baofeng radios worldwide because of their reliability and prices. These radios carry amazing features that one may seek to enhance further even though they perform well to date. Let's examine several ways to update and customize your Baofeng radio as well as the possible advantages of doing so.

1. Firmware Updates:

Performing routine firmware updates will greatly enhance your Baofeng radio's overall functionality and stability. Firmware updates are frequently released by manufacturers to fix faults, add new capabilities, and improve already-existing functionalities. Users may make sure their Baofeng radios are operating at their best and taking advantage of the newest features by routinely checking for and applying these firmware upgrades.

2. Antenna Upgrades:

You may improve the clarity and range of your communication by upgrading the antenna on your Baofeng radio. It will also enhance the radio transmission. The use of high-gain antennas or specialty antennas designed for specific frequency bands can yield better performance in areas with weak signals or high interference. To guarantee smooth integration and top performance, it's essential to choose antennas that work with your Baofeng radio.

3. Battery Enhancements: You may extend the life of your Baofeng radio by choosing high-capacity or extended-life batteries. This makes the radio perfect for extensive field operations or circumstances where power sources are hard to come by. The general dependability and longevity of your radio can also be improved by switching to higher-quality batteries, which will guarantee continuous operation and lessen the need for frequent recharging during important communication duties.

4. Audio changes:

You can enhance the audio clarity and output of your Baofeng radio by implementing audio changes, such as adding external speakers or amplifiers. Users can enjoy improved sound quality and clear, distortion-free communication even in demanding or noisy circumstances by integrating high-quality audio components. For customers who need accurate and dependable audio output during crucial communication activities, audio adjustments are very helpful.

5. Personalized Coding:

You can tailor your Baofeng radio to your unique communication requirements and tastes by customizing the programming. Users can tailor the radio's performance for different communication scenarios and operational environments by configuring unique channels, establishing preferred frequencies, and modifying the squelch levels. Custom programming improves overall communication efficacy and efficiency by allowing users to swiftly and easily access vital channels and streamline communication procedures.

6. Integration of External Accessories:

Including external accessories in your Baofeng radio, like carrying cases, headsets, and microphones, will improve its usability and functionality even more. Users may interact efficiently and pleasantly during a variety of activities and operations thanks to these attachments, which also enhance audio quality, allow hands-free conversation, and offer safe and easy transportation alternatives. If you want to get the maximum from your Baofeng radio, it is crucial to choose the proper and qualitative accessories.

Upgrades and Modifications: Upgrade/modifications may enhance and boost Baofeng radio's function and capability; however, such actions necessitate knowledge and caution, lest one incurs risk. Therefore, this means that you should follow manufacturers' guidelines on how to ensure its safety and optimal performance. Also, you should consult experts on which steps are safe for you to make without altering their functionality. It is equally important to appraise the specific communication needs and operating environment so that you can establish the most appropriate changes and improvements required within your own desires and tastes.

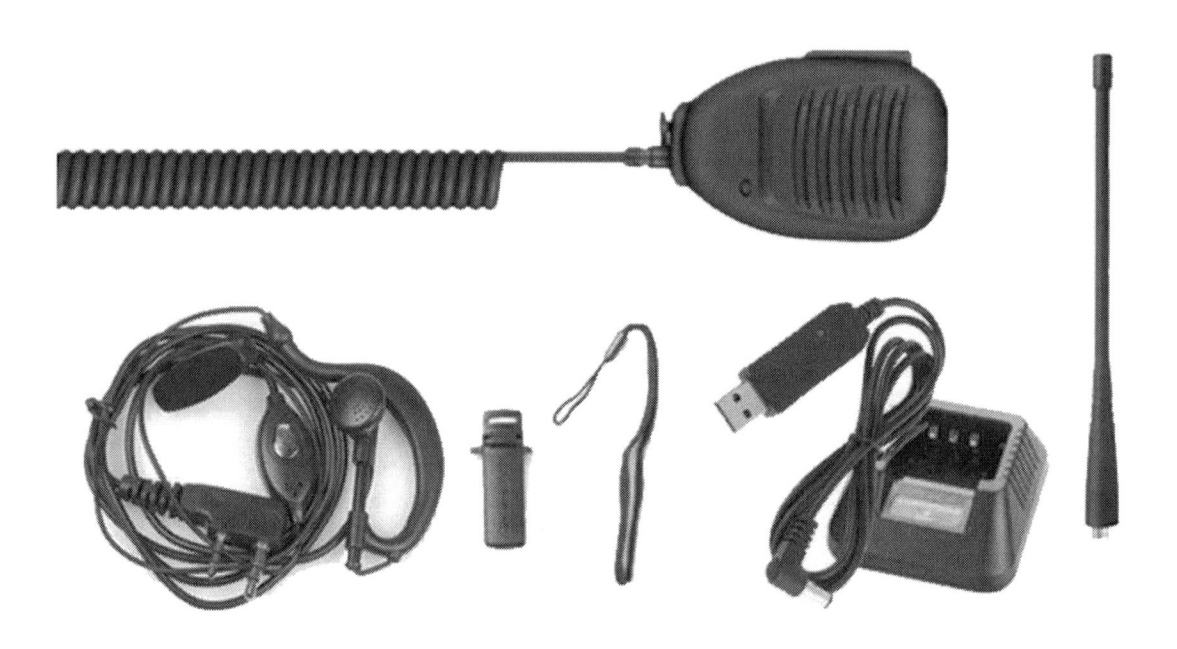

BOOK 5: RADIO ETIQUETTE AND COMMUNICATION PROTOCOLS

PROPER RADIO ETIQUETTE AND PRACTICES

Using the Baofeng UV-5R effectively and efficiently requires adhering to established communication protocols and using appropriate radio etiquette. Users can respect the requirements and priorities of other radio operators while maintaining clear and succinct communication by following these principles. The following are important guidelines to remember:

1. Take Up The Lingo

However, while using two-way radio communication, there are some essential aspects to keep in mind such as particular phrases and greetings to use when talking to another person or saying goodbye among others. It occurs because some of the ordinary phrases are not always converted into direct current or direct current, in other words, two-way radio waves. The use of a two-way radio like the Baofeng UV-5R resembles the secrecy in talking when a few words are cut off.

The following are a few of the most widely used walkie-talkie codes along with their definitions:

- Affirmative: Yes
- Negative: No
- Roger or Roger That: This indicates that the message was received and comprehended.
- Stand By: Please wait
- Over: I have finished speaking; often used at the end of a sentence to let the other party know they can speak
- Wilco: I will comply/follow instructions
- Copy or Read: usually used in a sentence to confirm your message was heard/understood, as in "Do you copy me?"
- Out: This is said to indicate the conversation is finished ("Over and out.")

Some of this jargon might be familiar to you from watching television shows or movies where people talk on two-way radios. Although learning lingo can take some time, it will guarantee that people can understand what you're saying.

2. Hold Off on Speaking

When you press the PTT, you need to pause for a while and give time to your words. This way, you will not be forced to repeat yourself in instances where the initial sentences in your audio or video have failed to capture the audience.

This way, you will not be forced to repeat yourself in instances where the initial sentences in your audio or video have failed to capture the audience.

Therefore, it is advisable that in such a situation should give your statement in full length so that none of your first four words will be cut out as that would make you say again what you have already said in advance.

3. Recognize Who You Are

That's why you must politely introduce yourself at the beginning of the chat despite the fact that most of the walkie-talkies do not have caller ID. Before saying your own name, you are supposed to state that of another person. Every employee in certain two-way radio-using industries has their own call sign. If this is applicable in your place of employment, make sure you address them by their correct call sign.

4. Keep Your Conversations Brief and Direct

When utilizing a two-way radio, try to avoid talking on the phone for extended periods of time. They were intended to provide fast communication bursts to address issues or complete tasks.

After one point says "break" and releases a button, it is necessary to give many instructions or cover a long way. It allows the other person an opportunity to talk if need be and then proceed to the next issue.

5. Learn The Phonetic Alphabet of NATO by Heart

Avoid using letters when you need to spell something out over a walkie-talkie because many of them sound the same. Instead, use the phonetic alphabet known as the NATO (North Atlantic Treaty Organization) to spell them out. To ensure clarity, this approach employs a term that is equivalent to each letter in the English alphabet.

To spell out a license plate that ends in EX, for instance, you would say these letters as "echo, x-ray" over a two-way radio.

6. Use a normal, clear tone of voice

Steer clear of talking too quickly when using a walkie-talkie. Talk normally; yelling or very quiet conversation might not be audible over the gadgets. To prevent your voice from seeming excessively loud to other radio users, place the radio's microphone three to five inches away from your mouth.

7. Don't Talk Over Other People

Wait until the other person's talk is over if you hear them using the two-way radios. Save your attempts to intervene for emergencies only. If you are conveying an emergency message, start your message by saying, "Break break break."

8. Assume That People Can Hear What You're Saying

When utilizing two-way radios, keep in mind that you do not have exclusive usage of the frequency and that anyone within hearing distance may be able to overhear your talks. Steer clear of sending any private or sensitive information in your communications until you are certain that your device is properly encrypted.

9. Speak in English Keep in mind that English is the official international language of two-way radio communication, unless instructed differently at work. You and your coworkers might have the necessary licenses to talk in many languages in particular situations.

10. Do Routine Inspections of Your Equipment

Therefore, it is advisable for all users of two-way radios to have pre-use equipment check tests to ascertain whether their batteries are fully charged while each radio is able to transmit clear messages. It could be necessary for your radio waves to circulate and, consequently, you should ensure that you are within reach of other people.

EFFECTIVE COMMUNICATION TECHNIQUES

Successful relationships are based on clear and effective communication, especially when using the Baofeng UV-5R. In order to guarantee smooth information sharing and maximum effectiveness in communication, take into account these essential methods:

1. Clarity and Precision in Messaging:

To communicate ideas clearly, speak intelligibly, and enunciate words accurately. Steer clear of confusing terminology or complicated jargon that could cause misunderstandings. To make sure that other users can understand your message, highlight the most important elements and communicate in a clear, succinct manner.

2. Active Listening and Acknowledgment:

Pay attention to incoming communications and quickly acknowledge those that you have received. This is an example of active listening. To demonstrate that you comprehend the information and are ready to respond appropriately, acknowledge the information you have been given. In order to facilitate efficient information flow and cooperative problem-solving, active listening creates a collaborative communication environment.

3. Standardized Communication Protocols:

To guarantee consistency and clarity in all contacts, familiarize yourself with industry-recognized terminology and standardized communication protocols. Respecting established communication standards reduces the possibility of misunderstandings or miscommunications during crucial operations and allows for easy coordination with other radio operators.

4. Prompt and Informed comments:

To ensure that communication flows continuously, answer incoming communications right away and provide prompt, insightful comments. Make sure your answers specifically address the issues brought up in the original communication in order to foster a positive discourse that facilitates effective problem-solving and decision-making.

5. Powerful Questioning Methods:

Use efficient questioning strategies to get accurate and thorough answers from other users. Ask questions that are clear and direct, with the goal of getting particular facts that will help you comprehend the situation more fully. Properly constructed questions encourage a more fruitful discussion and sharing of ideas among all involved parties.

6. Adaptability and Contextual Awareness:

Keep yourself flexible and sensitive to various communication circumstances, modifying your tone and style of speaking according to the topic at hand and the preferences of other radio operators.

Exhibit adaptability in your communication style, establishing a team atmosphere that promotes candid discussion and understanding between all members.

7. Effective Channel Management:

To avoid needless interference with ongoing transmissions, abide by specified communication channels and frequencies and follow established channel assignments. To guarantee that all users can communicate in an orderly and effective manner, abide by the frequency bands that have been assigned, and practice good channel management.

8. Building Collaborative Relationships:

Encourage an atmosphere of support and open communication that values constructive criticism and cooperation among all radio operators. Encourage teamwork and camaraderie among participants to facilitate the free flow of ideas and information that enhances the communication network's overall performance.

9. Emergency Communication Protocol Adherence:

Become familiar with the protocols and processes for emergency communication, with a focus on how important it is to prioritize emergency messages and follow established protocols as soon as possible in life-threatening situations. Be ready to respond quickly and effectively to emergency events by being ready to manage urgent scenarios and adhering to established communication procedures.

USING RADIOS IN DIFFERENT SCENARIOS (E.G., EMERGENCY, HIKING, EVENTS)

It is necessary to take certain considerations and adapt to the particular requirements of each context when using the Baofeng UV-5R in different scenarios, such as emergency circumstances, hiking adventures, or event management. This is a thorough guide that will help you use the Baofeng UV-5R in many situations:

Emergency Circumstances

Using the Baofeng UV-5R effectively can make a big difference in how quickly and efficiently responses are coordinated during emergencies.

Here's how to utilize the radio to its full potential in an emergency:

1. Guaranteeing Smooth Communication:

Make it a priority to perform routine radio checks and make sure the Baofeng UV-5R is always fully charged and functional to enable continuous contact during crucial situations.

2. Putting Emergency Procedures Into Action:

Respect established emergency communication protocols and procedures, placing special emphasis on using authorized emergency channels to quickly and efficiently transmit information to responding teams and necessary authorities.

3. Preserving Calm in Communication:

Communicate important information in a cool, collected manner, placing emphasis on precise and succinct messages to ensure that all parties are aware of the pertinent aspects.

4. Encouraging Coordination Efforts:

Work closely with emergency response teams and pertinent stakeholders to encourage effective and well-coordinated measures. This will guarantee that the Baofeng UV-5R continues to function as a dependable channel for crucial communication in times of emergency.

Trekking and Outdoor Adventures

Reliable communication devices, like the Baofeng UV-5R, are essential for traversing difficult terrain and navigating outdoor settings. Here's how to use the radio efficiently when going on hikes and outdoor adventures:

1. Signal Limitations Surveyed:

Learn about any possible signal restrictions in the hiking terrain and utilize appropriate techniques, like using additional antennas and greater power settings, to reduce signal obstructions and guarantee reliable connectivity.

2. Creating Explicit Communication Plans:

Before setting out on the hiking adventure, create detailed plans for communication that include agreed-upon frequencies and protocols. This will allow for easy information sharing and encourage group cohesion and safety.

3. Weather Conditions Monitoring:

Keep an eye out for shifting weather patterns and environmental factors, and modify communication plans as necessary to account for any unanticipated difficulties or dangers that may develop during the expedition.

4. Keeping in Touch with Other Hikers:

Make use of the Baofeng UV-5R to stay in constant contact with other hikers. This will guarantee that everyone stays in touch and is taken care of during the hiking trip, which will promote a cooperative and safe hiking experience for all participants.

Coordination and Management of Events

Effective communication techniques are essential for successful event management, and the Baofeng UV-5R can be a useful tool for promoting efficient coordination and information sharing during a range of events.

The following are some tips for making the most of radio use in event management scenarios:

1. Creating Clear Communication Plans:

To ensure a well-organized and planned approach to communication management during the event, create detailed communication plans that outline the precise roles and duties of each team member.

2. Creating Efficient Channel Assignment Plans: To avoid congestion and disruptions, create a clear channel assignment plan that will allow team members to communicate with each other without interruptions during the event.

3. Designating Team Leaders for Communication Oversight:

Assign competent team leaders to coordinate communication initiatives and efficiently convey crucial instructions. These leaders will play a crucial role in facilitating the maintenance of open and effective channels of communication during the event.

4. Regular Radio Checks:

To guarantee continuous connection throughout the event, conduct radio checks on a regular basis to confirm flawless connectivity and clear transmission. This will enable quick troubleshooting and modifications as needed.

5. Promoting Honest Communication Among Members of the Team:

Encourage a culture of open communication and cooperative problem-solving among team members. This will allow members to freely share ideas and information in order to quickly and efficiently handle any obstacles or changes to the original plan.

PRIVACY AND SECURITY CONCERNS

The biggest privacy and security issues with the Baofeng UV-5R are related to the possibility of data interception during transmission and unauthorized access to communication channels. Due to its open radio frequency design, the UV-5R presents a danger to the confidentiality of sensitive data since its transmissions could be intercepted and monitored by unauthorized parties. It is imperative to install encryption mechanisms and comply with regulatory criteria for authorized operation in order to offset these risks. Users may prevent security threats and preserve privacy with the Baofeng UV-5R by prioritizing secure communication routes and making sure that they comply with the law. Let's examine this radio's privacy and security issues:

Secrecy

Our first priority is your privacy. This policy illustrates how we collect, use, disclose, transmit to third parties, and use your personal information for different purposes. The reasons why we require personal information will be articulated while on the one hand; such information is being collected and thereafter.

We shall only collect and use the necessary personal information related to the said purpose and such other purposes unless authorized by the concerned party or under law.

We shall not retain your personal data any longer than necessary for accomplishing those objectives.

Collecting the personal data will be done legally and in fairways and, where the subject has given the knowledge and/or permission.

Personal data must be truthful, comprehensive, and in date, but also appropriate for the reason why these are processed or for which they are collected.

It will use prudent security measures to avoid losing or misappropriating as well as unauthorized revelation, reproduction, use, or change of personal data.

This will make it easy for customers to get information on how we handle/treat their personal data in accordance with our procedures and internal polities.

Policy for Application Privacy:

Location Information: No application on BTECH's end stores or shares your location. The program uses your location in order to share it with other application users. Besides switching on mobile and app location sharing, you also have to switch on mobile and application services.

Ensuring that our business adheres to these principles is done so as to safeguard the confidentiality of personal information.

Protection

1. Legality of Baofeng Radios

Compliance with FCC standards is necessary for Baofeng radios to remain licensed. While misuse of these radios can result in penalties, they are not intrinsically criminal.

Baofeng radios that use FRS, GMRS, or other approved frequencies have to be certified by the FCC and adhere to Part 95 regulations. It's against the law to use them on unapproved frequencies or with excessive power.

Baofeng's fame has also sparked worries about possible meddling and non-compliance. To guarantee the legal and responsible operation of Baofeng radios, it is essential to comprehend and abide by local communication legislation.

2. Regulations for Baofeng Radio

The FCC in the United States and equivalent authorities worldwide have regulations that Baofeng radios must follow. These rules guarantee correct frequency utilization, transmission power limitations, and interference avoidance.

If Baofeng radios are altered to operate outside of permitted frequency ranges or lack FCC certification, they may become unlawful and pose a danger of interfering with vital communication networks.

In order to prevent legal problems and maintain effective and secure wireless communication, users of Baofeng radios must operate them within the specified boundaries. While utilizing Baofeng radios or other communication equipment, it is essential to comprehend and abide by local radio legislation.

3. Baofeng Radio Regulations By The FCC

Radios made by Baofeng must abide by FCC rules in order to be used legally. They have to be properly certified, used only in permitted frequency ranges, and kept within power constraints.

In order to transmit on certain frequencies, users must have an amateur radio license. When Baofeng radios are used on unapproved frequencies or power levels, interference and FCC regulations may be broken, which may result in fines or other consequences.

It is imperative to comprehend and adhere to FCC regulations when utilizing Baofeng radios in order to guarantee lawful and conscientious communication.

4. Were Baofengs Declared Illegal by the FCC?

No, Baofeng radios are not now prohibited according to the FCC. Nevertheless, some models are prohibited because they may not comply with rules, they are not properly certified, or they use frequency ranges that are not approved.

In order to prevent legal problems, users must ensure legitimate operation by following license requirements and approved frequency usage.

5. Baofeng radios are banned: what's the deal?

Baofeng radios are prohibited because they may operate outside of approved frequencies if they are not properly certified by the FCC. These radios pose a concern to public safety because they can interfere with vital communication services.

In order to ensure proper use and prevent disruptions, regulatory bodies impose limitations on Baofeng radios, highlighting the importance of following communication guidelines and using radios responsibly.

6. How Safe Are Baofeng Radios?

When used sensibly and within the law, Baofeng radios are safe. However, there can be problems associated with their misuse, such as when they interfere with essential services or transmit at unlicensed frequencies.

The safe and efficient use of Baofeng radios for personal communication needs is ensured by following FCC laws, obtaining the necessary license, and operating within permitted frequencies, so as to limit any injury or disturbances.

7. Emissions of Electromagnetic Out of Baofeng

Like other electrical gadgets, Baofeng radios release electromagnetic radiation when they are in use. The potential for incorrect use or exceeding permitted frequencies to interfere with appliances and other communication systems is the cause for concern.

Any negative consequences of electromagnetic emissions are mitigated by adhering to FCC requirements, obtaining required licenses, and utilizing Baofeng radios responsibly. This ensures proper communication without interfering with or harming other systems.

TROUBLESHOOTING COMMON COMMUNICATION ISSUES

Baofeng UV-5R Won't Transmit

I apologize for the trouble you're having with your Baofeng UV-5R not broadcasting. Various factors can contribute to such a problem.

To try to pinpoint the problem, let's go through some troubleshooting procedures:

- Examine the battery. Make sure it is completely charged, or swap it out for one that is known to function. Low battery levels can occasionally have an impact on the radio's functionality.

- Make sure you are using the appropriate frequency and channel. Verify again that the radio is in the proper mode (memory or VFO) and that you are using the correct frequency.

- CTCSS/DCS Codes: Make sure that the sending and receiving radios are using the same set of codes if you are using CTCSS or DCS codes. The radios won't talk to each other if they don't match.

- Monitor Function: Verify whether or not the "Monitor" function is active. It might stop transmission if it's turned on. Please switch it off and try sending it once more.

- Squelch Level: If the squelch level is set too high, transmissions may be blocked. To see if it helps, try reducing the squelch level.

- Voice-Activated Transmission, or VOX: If this function is turned on, the radio may only start talking when it hears sound. Turn off VOX and observe any changes.

- Antenna: Verify that the radio's antenna is firmly fixed. Transmission may be impacted by a damaged or loose antenna.

- Interference: Look for any nearby sources of interference that might be affecting the radio's operation, like other electronic gadgets.

- Reset: To rule out problems connected to software, try factory resetting the Baofeng UV-5R.

- Legal Frequency and Maximum Power: Make sure that the radio is being used within the permitted power and frequency ranges. Transmitting data in unapproved frequencies or with too much power can cause problems.

Should you have attempted each of these solutions and the problem still exists, there may be a hardware issue with the radio. It is advisable to get in touch with Baofeng customer service or a qualified radio technician in that situation. They are better able to identify the issue and offer a fix.

BOOK 6: BAOFENG RADIO MAINTENANCE AND CARE

CLEANING AND MAINTENANCE TIPS

The Baofeng UV-5R is a flexible and durable handheld radio, noted for its durability and functionality. To maintain its top performance and extend its longevity, regular cleaning, and adequate maintenance are needed. Here's a thorough guide to help you keep your Baofeng UV-5R in peak condition:

1. Dust and Debris Removal:

Regularly dust the outside surfaces of the radio using a soft, dry cloth. Pay great attention to the crevices and buttons to ensure that no dirt builds over time. Avoid using abrasive materials or harsh chemicals that could potentially damage the device's coating.

2. Screen Cleaning:

Use a microfiber cloth to gently wipe the screen and remove smudges or fingerprints. If required, lightly dampen the cloth with water or a mild glass cleaner, ensuring that no liquid penetrates into the device. Carefully dry the screen to avoid any moisture from causing damage.

3. Antenna Maintenance:

Inspect the antenna routinely for any signs of wear or damage. Wipe off the dirt and dust from the antenna using a moist piece of cloth. Make sure that the antenna is tight and in good condition because it is important in receiving and transmitting signals.

4. Battery Care:

Maintain clean battery contacts by routinely examining and wiping them with a soft cloth. If the contacts appear dusty or rusted, use a cotton swab lightly soaked with rubbing alcohol to clean them carefully. Properly store and charge the battery in a cool and dry area to maintain its longevity and overall performance.

5. Safe Storage Practices:

Store the Baofeng UV-5R in a dry and cold environment to prevent moisture or humidity from harming its internal components. Consider using a protective case or cover to shield the gadget from dust and any physical damage when not in use. Avoid keeping the radio in direct sunlight or severe temperatures, as these conditions might harm the gadget over time.

6. Firmware upgrades:

Regularly check for and install firmware upgrades offered by the manufacturer. Firmware updates are critical for maintaining the radio's best functionality since they generally include performance advancements, bug fixes, and new features that improve overall user experience and communication capabilities.

7. Proper Handling Techniques:

Handle the Baofeng UV-5R with care to prevent accidental drops or impacts that may cause internal damage. Use a secure and comfortable hold when operating the instrument, limiting the danger of unintended misuse or potential physical harm to the radio.

8. Routine Inspections:

Conduct routine inspections of the device, including the buttons, knobs, and connectors, to ensure that all components are in proper operating condition. Check for any loose parts or anomalies that may influence the radio's functioning, and solve any concerns promptly to prevent future damage or degradation.

9. Preventive Measures in Harsh Conditions:

Protect the Baofeng UV-5R from severe temperatures, direct sunlight, rain, and other harsh environmental conditions that could potentially affect its performance. Implement preventive steps, such as utilizing protective coverings or maintaining the device in a climate-controlled area, to limit the impact of adverse conditions on the radio's components.

BATTERY MANAGEMENT AND CHARGING

If you have access to a 120v outlet, charging your Baofeng UV-5R platform radios (BF-B8HP, UV-5RX3, etc.) is easy, but using a handheld ham radio that way is uninteresting. As the primary function of a handheld transceiver, sometimes known as a handy-talkie, is to facilitate communication while on the go, I wanted to be sure I could maintain radio functionality under a variety of conditions. To help me get the most out of my radios, I cover three batteries in this part along with a few charging solutions that I use on a regular basis. These possibilities include using an aftermarket battery pack with standard AA batteries, charging in your car, and using solar panels when trekking.

Type 1: Only Cradle-Charged

The quickest method for charging the batteries in this article is using the charging cradle that comes with the UV-5R, which plugs into a standard household socket. While two of the three batteries—the 1800mAh and 2100mAh that come with different UV-5R platform models—can only be charged using the cradle, the other two are compatible with the charger. This reduces mobile usage possibilities.

You may still use the charging cradle while on the go, though. A 10VDC input—a volt direct current, similar to that of a battery—is needed for the cradle. Baofeng offers a Baofeng Transformer Cable that may be used to modify the voltage on USB outlets to make them compatible with the charging cradle, even though the radios do not come with one. With the help of this cable, you may power the cradle from any device that has a USB connection, such as a computer, emergency auto starter, hiking solar panel charging kit, or cigarette lighter converter. You can multitask elements in this way to make sure your kit functions as a cohesive unit. For instance, you can charge your radio, headlights, cell phone, and other devices by using the emergency car starter.

To give you an idea of the bulk if you decide to go with this option, the first picture below shows both batteries together with the wall charger, charging cradle, and aftermarket transformer wire. In the event that you wish to construct your own charger, the second image just displays the charger's electrical specifications. (DIY/MYOG/Use caution!)

The 1800mAh and 2100mAh batteries have different thicknesses when they are put into the radios. There's a small change if you look attentively at the top corner of the battery, where the belt clip is attached. The 2100mAh will provide you with a little longer battery life without sacrificing much size if you plan to buy extra batteries to carry along and you don't want the Extended Battery. You will notice a slight change in the clip, but it won't be noticeable unless you are attempting to hang the radio on an extremely thick object.

The transformer cable and charging cradle combination offers a cost-effective solution for adding portable charging capabilities.

An additional benefit is that, in contrast to the Extended Battery or AA Battery Pack, utilizing the smaller batteries keeps the radio small. This matters for some carry alternatives. For instance, the 3800mAh battery does not fit inside the EMDOM MM GPS Pouch, therefore I use that instead when I wish to place the radio on a belt or backpack. The radio and the 1800mAh and 2100mAh batteries both fit inside. (A shoulder or lapel mic is needed for this; I use the Code Red Signal 21-K.)

However, it has two main drawbacks. Kit size as a whole comes first. The cradle is bigger than the extended battery with its own USB cable, which is covered below, and you will need to bring it with you if you want mobile charging alternatives. (The USB cords cannot be used together.) However, this is a nice alternative if you're carrying the cradle anyhow, such as if you're heading to a set site with grid power and merely want a backup.

The second drawback is the speed at which it charges; using this option seems to be slower than putting it in the wall and far slower than using the extended battery option, which does away with the cradle.

If you don't want to waste the OEM batteries, get some extended batteries when building the entry-level EMCOMM kit. Additionally, I put them in some of the kits because they can only be charged with the cradle. One method I employ to maintain organization is to place the transformer wire and wall charger cradle within a zipper lock storage bag, and then store the USB charger for the extended battery inside the designated zipper lock bag. I've already removed the complication of trying to figure out which pieces work in an emergency if an untrained family member needs to use and charge the radios.

Long-lasting (3800mAh) battery with USB charging cable

The best compromise, in my opinion, is the Extended Battery (3800mAh) and USB Charging Cable. It fits a longer battery life into a lot smaller package—far smaller than lugging the cradle along—while having a considerably higher capacity. For quicker charging, it works with the charging cradle when grid power is available, but it can also be used with any USB outlet without the bulk of the cradle. This translates to a more compact total kit size while maintaining the same capacity to use portable solar power sources, car chargers, emergency starters, and tiny USB battery packs.

The size of the Extended Battery is really its main drawback. It increases the radio's height, which may reduce carry alternatives. The radio will not fit with the 3800mAh battery fitted, despite the EMDOM Medium GPS Pouch's great convenience, as I have already mentioned.

The Extended Battery and its charging choices may be seen in the first picture below. It can use the transformer cable, wall charging cradle, or special USB cable. If you can use the USB cord alone and leave the charging cradle at home, you'll see a difference in the total size of the equipment.

An intriguing USB cable side effect is depicted in the second picture. You can get a ballpark idea of the battery's charge by plugging the LED into the battery rather than a power source. It is also very bright as well as a nearly fully-charged battery. The advantage of such a design is that since it does not need to be attached to the radio, one may re-charge a second battery already in use and therefore have such an option available if your power outlet fails you! Moreover, you may also conduct this operation using a portable battery charger which you take with you when you are traveling and even put in your bag. However, take care to control the heat produced during charging. If at all possible, I'd advise placing it in an external pocket with ventilation.)

The third image only illustrates that the battery can be charged on the cradle apart from the radio. This is what all three batteries are able to do.

Two AA batteries

I have an AA battery pack in every car kit I own because it's a wonderful emergency alternative. Five AA batteries are required to power the radio (along with a dummy battery that is included). This implies that you might be able to run your radio with items you find at the grocery store or gas station even if your Baofeng battery is dead. Alternatively, since I see my entire kit as a system that should function as a complete rather than the communications kit as a stand-alone item, I already own several electronic devices that require AA batteries, as well as a portable GoalZero AA recharger. With the GoalZero charger, I can now use standard AA batteries to power my phone, radio, headlamp, and other devices. However, I usually pack rechargeable AA batteries with it so that I can use the GoalZero charger to refuel the batteries and then use them to power whichever gadget I'm focusing on at the moment.

Take note of the warning mentioned in the Amazon reviews regarding the possibility that a piece could break away and pose a fire danger. Reviewers claim that while it isn't on every box, it can be readily rectified when it is.

"Eliminator" battery.

Although it allows the radio to utilize the vehicle's battery (via the cigarette lighter plug) instead of the radio's battery, the BL-5 Battery Eliminator is a bit of a misnomer. It doesn't actually remove the need for a battery. This might be perfect in a lot of scenarios, like letting you save your power when driving and just replace it when you're walking. But, when you use it, you remain attached to the vehicle, and drawing electricity from the car may interfere with the radio's operation.

But, when you use it, you remain attached to the vehicle, and drawing electricity from the car may interfere with the radio's operation.

While you have it, you are stuck in the car with it. Electricity in the car can also interfere with the radio.

However, it seems like a reasonable option under certain circumstances.

In fact, I buy at least one of each kit I possess and use it when traveling long distances in a car with no mobile device available. It's convenient because I will never have to think of leaving the HT turned on as I exit my car and neither do I bother to keep recharging the batteries of the device.

It simply goes on when the 12-volt gets power, which is when I start the car, and it shuts off when that happens.

STORING AND PROTECTING YOUR RADIO

Maintaining the longevity and peak functionality of your Baofeng radio requires proper storage and protection. Take into account the following advice to protect your device:

1. Use a protective case: To protect your Baofeng UV-5R from scratches and other external factors, get a sturdy carrying cover or holster made especially for it.

2. Keep it dry: Store your radio in a dry place to avoid moisture damage. To lower humidity, think about utilizing a moisture-absorbing material in the storage location, like silica gel packs.

3. Steer clear of extreme temperatures: The performance and battery life of your device may be impacted by extremely high or low temperatures. When at all possible, keep it in a temperature-controlled space.

4. Routine maintenance: Look for wear, corrosion, or damage on your Baofeng radio on a regular basis. Wipe off any dust or debris using gentle cloth from the exterior, and ensure that all of its buttons and knobs are in good working condition.

5. Secure storage area: Ensure that you maintain your Baofeng radio in a safe location so as to minimize the risk of it being stolen or accessed illegally. When not in use, think about securing it in a drawer or cupboard.

FIRMWARE UPDATES AND SOFTWARE MANAGEMENT

It's important to comprehend the relevance of firmware upgrades and software management before diving into their specifics. The Baofeng UV-5R's firmware is the core software that manages its hardware. It controls important features and operations, guaranteeing smooth functioning and performance. On the other side, software management entails managing the device's configuration, programming, and settings. Users can adjust the channels, frequencies, and other variables to fit their needs and tastes by adjusting the software.

Looking for Updates to the Firmware

Updating the firmware on a regular basis is essential to keeping the security and functionality of the Baofeng UV-5R intact. Firmware updates are usually released by Baofeng to fix bugs, enhance functionality, and add new features. Use these procedures to see whether there are any firmware updates:

Go to the Official Website (1) Go to the official Baofeng website to get started. Go to the downloads or support department, where firmware updates for the UV-5R are probably available.

2. Check for Harmony: Make sure the firmware update is appropriate for the particular Baofeng UV-5R model you own. Device issues and possible harm might result from installing mismatched firmware.

3. Get the most recent firmware update: After you've verified compatibility, download the computer's firmware update. Make sure you store it somewhere you can quickly access it in the future.

Changing the Firmware

Your Baofeng UV-5R's firmware update is an important procedure that requires close attention to detail. A mistake made during this process could cause the gadget to sustain irreversible damage. Here's how to upgrade the firmware step-by-step:

1. Connect the Radio to the Computer: Use an appropriate programming cable to connect your Baofong UV-5R radio to a computer.

Throughout the update process, make sure the connection is reliable and secure.

2. Install Firmware upgrade Software: You must install the Baofeng software before you can upgrade the firmware. Refer to the instructions accompanying your software program to ensure the accuracy of the installation process.

3. Backup Settings: You should first make a backup of your current settings before proceeding with the firmware update. The fact of this safety measure is that in case you face some difficulties during the upgrade you will be able easily to switch back to the previous settings.

4. Start the Update: Begin with launching the firmware update software onto your computer and follow the instructions displayed on the screen. Do not hurry, otherwise, you risk doing more damage to your computer that can be undone only in a different world.

5. Check the Update: Once complete, be sure that the firmware is properly set. Confirm that the settings have not been altered and check if operations are running as expected. If there are any differences, refer to the user manual or contact Baofeng's customer service for assistance.

Software Management The Baofeng UV-5R's software must be managed in addition to firmware updates in order to preserve customized settings and configurations. Through effective software management, users can customize the gadget to meet their individual requirements. This is a thorough rundown of the main elements of software management:

1. Channel Programming: To program and manage channels in accordance with your preferences, use the Baofeng software. To simplify your communication process, group channels according to geography, frequency, or any other pertinent factors.

2. Frequency Customization: Adjust the frequencies to suit your needs for usage. Make sure the frequencies programmed adhere to the rules set forth by the appropriate authorities and are not outside of the law.

3. Customization of Settings: Optimize the device's performance and usability by adjusting the settings. Adjust variables including display preferences, squelch settings, and audio volumes to suit your communication requirements.

4. Restore and Backup: Make regular backups of your program configurations to protect against unanticipated data loss or firmware problems. Having a backup guarantees that in the event of any errors or conflicts throughout the programming process, you can promptly restore the configurations.

The Best Methods for Software Management and Firmware Updates

Follow these recommended procedures to guarantee a seamless and trouble-free firmware update and software administration experience for the Baofeng UV-5R:

1. Study and Preparation: Do an extensive study and compile all relevant data before starting any updates or modifications. Recognize the effects of the upgrade or any programming changes to foresee any difficulties.

2. Pay Close Attention to Instructions: Pay close attention to the guidelines that Baofeng provides when updating firmware or managing software. Disregarding the instructions may do the gadget irreparable harm.

3. Frequently Backup Your Data: Establish a routine of often backing up your data and settings to reduce the dangers related to software modification and firmware updates. Keeping your backups current guarantees that, in the event of a problem, you can promptly restore the configurations.

4. Seek Expert Guidance When Required: If you run into difficulties or doubts when updating the firmware or managing the software, get help from seasoned experts or ask Baofeng's customer service for advice. Refrain from taking needless chances that can jeopardize your device's functionality.

BOOK 7: ADVANCED BAOFENG RADIO FEATURES

GPS AND APRS INTEGRATION

So... What is APRS? APRS stands for Automatic Packet Reporting System. Amateur Radio Operators (Hams) use APRS for many different things. In a nutshell, APRS lets you transmit your GPS location out onto the radio net for others to pick up and place you on a map. This is the basic function. Other functions include messaging, relaying, I-Gating, and Weather (I'll touch on these parts later).

What is a UV-5R? It is a great entry-level, hand-held radio. Inexpensive and forgiving for those new to the hobby.

Why am I putting this together? Well, to share the knowledge and put all the research I found into one spot.

I'd like to call out to KF7BBI (Dave) and KG7IOE (Terrance) for helping me troubleshoot the issues I was having. And KC2UHB (Diana) just because.

How do I use APRS? I like to snowshoe in the winter and it is one way I can let others know where I am in case of an emergency.

STEP 1: Things You Need for APRS

Three basic needs (plus one requirement)

- A radio
- A cable
- An APRS solution. Mine is a smartphone with APRSDroid - https://aprsdroid.org/

(The requirement is having an Amateur Radio License)

Description:

My trusty UV-5R with extended life battery and upgraded antenna (check Amazon or eBayz for the cool dealz)

This cable was a great find by Dave. It provides ground isolation so that the PTT is activated on transmit. If you don't have this (or a TNC) the radio will key open and disturb others (see Ham

Etiquette). The cable was put together here: https://github.com/johnboiles/BaofengUV5R-TRRS (another call out to John Boiles for this awesome project).

My phone is a previously used Samsung Android-based smartphone. The key to this is to factory reset the phone and not log it back into Google... more on that later.

To operate (that means transmit) on Amateur Radio frequencies, one must get a license. The best way is to search around your community for a local club that can guide you through the study material for the test. 'Technician' is the first level of license and is all you need to Transmit. If you want to monitor or listen, no license is needed... but what fun would that be? I've had mine for 3 years now... KG7IOA.

STEP 2: Configuration

It all comes down to the details!!! This is were Terrance really helped out.

This configuration is how I was able to put all the parts together and make it work. I did struggle as my transmissions were not hitting the relay or being rebroadcast to the I-Gate. Terrance had a known-good working setup on his phone that I was able to test with and compare configurations to my phone to get it to work.

What was happening ... well when APRSDroid would transmit, I would see the transmission from my radio but I wouldn't see it come back from the Relay or hit the I-Gate. Turns out that the default setting in APRSDroid is to send the signal out on the 'phone/voice' of the phone. This signal was so incredibly weak that I couldn't get the relay to pick it up when I drove past it.

I switched to the 'Ringtone' setting in APRSDroid. This kept sufficient signal power for the transmission to be picked up by the Relay which passed it onto the I-Gate.

The Vox on my radio was an enigma wrapped in a mystery as well... the radio documentation (and much of the Interwebz) couldn't really tell me that Vox setting 1 was open-all-the-way sensitivity and 10 was just-slightly-open sensitivity. I only found this by trial and error on my own while watching my signal being pushed out from my radio. The radio has an LED that illuminates, to the color of your choice, on Tx and Rx (transmit and receive).

Update!! v2

Phone volume must be 3/4 of the bar. You will hear the 'squawk' of the transmission but if the volume is too low, even with Vox at 1, nothing will transmit. I put a sound muffler on my phone speaker so it doesn't spook me while driving.

These are the settings that worked best for me to get this whole thing working properly... your mileage may vary.

Update v3

See the updated photo of the volume settings... This will be completely silent, sending all AFSK out the headphone jack to the radio.

STEP 3: Conclusions...

Filling in the gaps...

Why did I not log my phone into Google? Once you do, all the notifications from Insta-Face, Snap-Books, and the like will come through the phone when one of your compadres posts a new cat video or a new noodle truck. And if you are APRSing, those notifications will then be transmitted over the air. "Das ist verboten" per the us FCC rules for spurious transmissions. Also, I don't have a SIM card in the phone.. who wants to get a phone call out in the mountains? That's the whole reason for getting out and about.

What is a Relay and what is an I-Gate? A Relay is just that... it takes your transmission, usually of lower power, and rebroadcasts at a higher power so that it covers a larger area... Search and Rescue can't get to you if your coordinates aren't pushed out.

APRSDroid takes advantage of the GPS of the smartphone for coordinates. That is what gets transmitted along with your callsign and optional short message.

APRSDroid also allows for direct messaging to individuals... just slide over to the "Messaging" tab to send a note to another Ham using their callsign. They can then ack from their APRSDroid.

I didn't forget... an I-Gate is a relay that is connected to the Interwebz. Why? A very cool Fin put aprs.fi together to take APRS transmissions and post them onto a modified Google Maps (uses the Maps API). That way you can see the Hams in your area or those traveling through or if you want to send a message to a fellow Ham in Australia or Scotland, you can see if they are out and about with APRS capability.

(Pro Tip: If you are geographically separated from one of your compadres [not in LOS between the radios], a message can be sent through an I-Gate similar to IRLP. This is untested by me, though).

Oh and Weather!!! One of my other hobbies... APRS can be used to transmit weather data from compatible weather stations... why? For the fun of it!!! The APRS Wx packets are simple enough to program into your 'duino/bone and send over the link. I have an additional project that I am ruminating on to inject Wx data into my APRS tracking while out snowshoeing.

Finally, if you are into hard-core 'tactical reclamation operations' (e.g. hacker, et al.) this chap (http://unsigned.io/projects/microaprs/) has a home-brew version that could easily accommodate a GPS shield to his 'duino project.

CROSS-BAND REPEATING

To communicate, cross-band repeating means getting messages on one frequency and sending them at the same time on a different frequency. This feature makes it possible to communicate over longer distances and better through barriers like buildings, mountains, or thick vegetation. Users can improve the quality of their conversation by using cross-band repeating. This is especially useful when two devices on the same frequency can't directly talk to each other.

Useful Applications

The Baofeng UV-5R's ability to repeat across bands can be used in a lot of different situations, such as:

Cross-band repeating can be very helpful in emergency scenarios where direct communication might not be possible, like during natural disasters or search and rescue operations. By strategically placing one radio in a place with better signal reception and the other in the middle or up high, contact can be sent over longer distances, making sure that rescue efforts are coordinated and that important information gets to everyone.

Communication in Rural places: Cross-band repeating can help people talk to each other in rural or remote places that don't have a lot of phone lines. It makes reliable contact possible between people or groups that are located far apart, so they can stay in touch and plan things like farming, hunting, or fun activities that they can do outside.

Radio Operations for Amateurs:

People who are into amateur radio often use cross-band repeating to improve their ability to talk to each other during contests, trips, or events. Cross-band repeating gives operators a longer range, which lets them set up communication links over long distances. This lets them connect with other operators in the same area or even around the world.

Meetings and events open to the public:

Communicating with event organizers, security personnel, and people attending during public affairs, festivals, and gatherings will help ensure the smooth running of an event and address challenges that arise. Cross-band repetition helps build a strong communication network that covers the whole event area. This makes it easy to share information and act quickly if something unexpected happens.

How to Set Up Baofeng UV-5R Cross-Band Repeat

Setting up the Baofeng UV-5R so it functions ideally is necessary for successful cross-band looping. For a full explanation of how to set up cross-band repeats, see this:

Step 1: Pick out frequencies

Pick out the bands that you want to send and receive. Ensure the frequencies conform to what the relevant regulatory bodies have sanctioned. Select frequencies that suit your contacts but watch out for interference and traffic on those frequencies near you.

Step 2: Turn on Cross-Band Repeating Mode.

On the Baofeng UV-5R, go to the menu and find the choices for cross-band repeating mode. The cross-band repetition should be switched on, and the frequency, offset, and tone should be dialed accordingly. Ensure that you adhere to the user manual's instructions on how to correctly set up and activate this function.

Step 3: Set up the antenna

Set up antennas that are right for both the sending and receiving bands. Use appropriate antennas of good quality to maximize results. Proper installation of antennas facilitates clear transmission and receiving of sounds without interference that results from having conflicting communication frequencies.

Step 4: Make sure the cross-band repeating works.

It is important to try the cross-band repeating feature thoroughly to make sure it works and find any problems or performance limits. To see how well the setup works in real life, test it in a variety of settings, such as ones with different amounts of interference, obstructions, and distances.

Step 5: Adjust the settings for the best results

Based on the test results, change the settings to get the best performance from the cross-band repetition. You can fine-tune things like signal strength, audio quality, and sensitivity to make sure that contact is clear and reliable across the chosen frequencies. Make the changes that are needed to keep interference to a minimum and get the most out of the cross-band repeater setup as a whole.

Things to think about for the best performance

When using the Baofeng UV-5R's cross-band repeating function, it's important to keep a few things in mind to get the best performance and communication:

Planning for frequency:

Ensure that the frequencies have been planned carefully so as not to interfere with or clash with other radio systems in the area. Research what frequencies are permitted where you live, and pick out those that give you maximum communication with minimum disturbance.

Picking an antenna:

To make sure that signals are sent and received efficiently, choose high-quality antennas that are made for the VHF and UHF bands. To get the most out of the cross-band repeater setup, think about things like antenna gain, polarization, and radiation pattern.

How to Manage Power:

Controlling how much power the Baofeng UV-5R uses is important for keeping transmission stable and extending the battery life. Change the power levels based on the communication range and the environment to find the best mix between saving power and sending signals reliably.

Compliance with regulations:

Following the rules and instructions set by the appropriate regulatory bodies for radio communication and frequency usage is important. Make sure that the cross-band repeating setup follows the rules for power limits, frequency assignments, and any other legal requirements to stay out of trouble with the law and keep other radio systems from working too closely.

VOICE-ACTIVATED OPERATION

The fact is that the VOX functionality is enabled in certain types of Baofeng UV-5R handheld units, which allows for the immediate detection of any kind of sound or speaking signal, leading to an automatic transmission even if the PTT button is not pressed. This feature helps those individuals who require talking on phones without being distracted when doing a task that needs their hands. The physical ptt button gets replaced by the VOX communication tool that enhances communication, user convenience, and efficiency.

How the Baofeng UV-5R's VOX feature works technically

The Baofeng UV-5R has a number of technical features that make the VOX tool work well:

Levels of Sensitivity:

The UV-5R has sensitivity levels that can be changed, so users can change how the VOX feature responds to different levels of background noise and sound volumes. By changing the sensitivity levels, users can make the device better at picking up voice signals and less likely to give false alarms because of background noise or other environmental issues.

Delay Settings: The device lets users set delay settings, which determine how long the communication will continue after the voice input stops. Users can fine-tune the VOX feature to meet their specific communication needs and preferences by changing the delay settings. This makes sure that transmission management during conversations runs smoothly and efficiently.

Headsets and microphones that work with it:

The Baofeng UV-5R works with a number of headsets and external mics that support the VOX feature. If a user has compatible accessories, they can connect them to the device. This lets them talk and respond without using their hands. This makes the UV-5R more useful in a wider range of contact settings and situations.

Power Management: For the Baofeng UV-5R to keep working well during VOX usage, it needs to have good power management. The gadget has power-saving features that make the best use of energy while still allowing voice detection and transmission to happen continuously. By handling power well, the UV-5R makes sure that the batteries last longer, so they can be used for longer without having to be charged or replaced often.

How the Baofeng UV-5R's VOX feature can be used in real life

The Baofeng UV-5R's VOX feature can be used in a number of personal, recreational, and business settings, such as, but not limited to:

Hands-Free Communication While Driving: With VOX, drivers' hands are never off the wheel nor do their eyes go out of the windows whenever they use cell phones while riding in a car.

With the VOX feature and compatible headsets or microphones, drivers can easily talk to each other on long trips, in emergencies, or during convoy operations. This allows for safe and effective conversation while keeping their attention on the road.

Better coordination in outdoor activities: VOX makes it easy for people to talk to each other without using their hands while hiking, camping, or playing leisure sports. Users can stay in touch and plan activities without having to physically change the radio settings. This makes sure that communication flows smoothly and improves situational awareness in outdoor settings that change quickly.

Streamlined Work in Business Environments:

VOX makes it easier to communicate and improves operational efficiency in professional settings like security, event management, and building sites. While doing their jobs, security staff, event coordinators, and building teams can talk to each other without any problems. This lets them make quick decisions and respond quickly to changing situations without any delays or interruptions.

How to Make Specialized Environments Accessible:

VOX is a great way to communicate without touching the radio device. This is especially useful in places like hospitals, workplaces, and labs where people may not be very good at using their hands or where it is important to keep things clean. Healthcare workers, factory workers, and lab technicians can all speak clearly while following safety rules and keeping operations running smoothly.

Setting up VOX on a Baofeng UV-5R

Setting up and configuring the Baofeng UV-5R's VOX feature requires a few important steps to make sure it works right and performs at its best.

Here is a full guide on how to set up and use VOX on the Baofeng UV-5R:

Step 1: Go to the device's settings

Baofeng UV-5R's control panel will lead you to the setting menu. Find the 'VOX settings' submenu and then click it to adjust the various sensitivity levels, delay settings, and external device compatibilities.

Step 2: Change the levels of sensitivity

Change the levels of sensitivity based on the surroundings and how you plan to use the VOX feature. Fine-tune the sensitivity levels to make sure that voice recognition works correctly and that background noise or other sound sources don't cause too many false triggers.

Step 3: Set up the delay settings

Choose the right delay length to control how the communication continues after voice input stops. Change the delay settings to fit the users' needs and tastes for communication, taking into account things like the speed of the conversation, the time it takes to respond, and the communication protocols being used.

Step 4: Connect any accessories that are compatible

By connecting headsets or microphones that are compatible with the Baofeng UV-5R, you can use the VOX option to talk without using your hands. It is important to make sure that the accessories are firmly connected and set up so that voice detection and transmission work without any problems.

Step 5: Make sure the VOX works.

Test VOX extensively in multiple scenarios, just to ensure that it is operational and effective. Evaluate the effectiveness of sending voice messages via the VOX function as the device is exposed to varying levels of background noise and voice volumes.

Step 6: Choose "the performance" as the highest setting.

Make necessary adjustments to the VOX settings using the output of the test in order to enhance the functionality of the feature for reliable, fast conversations. To make hands-free conversation with the Baofeng UV-5R work smoothly and easily, make the necessary changes to the sensitivity levels, delay settings, and accessory configurations.

Things to think about for the best VOX performance

To make sure that the Baofeng UV-5R works well and that the VOX function is used correctly, keep the following important things in mind:

Levels of background noise:

When setting the VOX feature's sensitive levels, you should take into account the noise level in the area where it will be used. Change the sensitivity levels to get rid of background noise while still receiving and sending voice signals correctly, which will improve the communication experience as a whole.

Training and getting to know the user:

Ensure that individuals receive adequate training and time to accustom themselves to the functionality and various components of the VOX feature. Instruct people on the correct usage of VOX with details on setup, configuration, and optimal use, in order for it to work perfectly and conversation flow well.

Accessories that work with it:

Make sure the Baofeng UV-5R works with the headsets or microphones you connect it to so that the merging goes smoothly and the performance is reliable. Choose high-quality extras that are made to work with the Baofeng UV-5R and its VOX feature. Choose reliable headsets or mics that work well with the VOX feature and don't get in the way of other features when you're talking on the phone without using your hands.

Things in the environment:

Think about things in the surroundings, like temperature, humidity, and physical obstacles, that could affect how well the VOX feature works. Making sure the Baofeng UV-5R is well protected from bad weather is important to keep it working properly and extend its life, which will ensure stable VOX performance in different settings.

Care and maintenance:

Do regular maintenance and care on the Baofeng UV-5R and its VOX feature to keep them working at their best. Check, clean, and update the software on the device on a regular basis to avoid technical problems, extend its life, and make sure it works well for conversation while it's being used.

Getting along with the rules:

Follow the rules and guidelines set by the government for radio contact and device use to stay within the law and avoid interfering with other communication systems. Learn about the rules and laws that apply so that you can use the Baofeng UV-5R correctly and avoid breaking them, which could result in legal problems or problems with how the business runs.

Getting the Most Out of the Baofeng UV-5R's VOX Function

Utilizing the Baofeng UV-5R's VOX feature can improve users' conversation experiences and speed up their work processes in a variety of applications. Some of the best reasons to use the VOX tool are:

Better Convenience: The VOX feature has a hands-free option that facilitates user interaction without physically operating the device. The convenience aspect is particularly crucial in busy environments such as these because it simplifies good communication and efficient workflow in terms of doing more than one thing at a time, as well as being on the move.

Improved Safety: VOX enables people to talk with their hands in hazardous areas or situations. Voice-activated communication on the Baofeng UV-5R lets users stay focused on important tasks or emergencies while keeping connected and up to date.

Efficient Coordination: The VOX feature simplifies communication between team members, colleagues, and people in groups. Collaboration is thus enhanced and any decision-making process becomes a collective effort. The VOX capability eliminates hindrances and interruptions that disrupt communication, facilitating the sharing of information and ease of work among many businesses and individuals.

Flexible Use: The versatile VOX feature is applicable under various scenarios, including outdoor expeditions, work assignments, recreational activities, and emergencies. Adaptable and free, this makes it an effective means for people and businesses that need to communicate effectively based on diverse preferences among a large number of audiences.

.EXTERNAL DATA CONNECTIONS (E.G., APRS)

One popular example of a handheld transceiver is the Baofeng UV-5R, which supports various external data links including APRS and other functionalities.

1. The automation packet reporting system, commonly referred to as APRS, is a digital approach for hams to transmit data in real-time, such as GPS data, weather reports, and SMS messages. The Baofeng UV-5R can be linked to an outside unit or module to enable APRS functions. This lets users share information about where they are and other important details.

2. VOX (Voice-Operated Exchange): This function lets you use the phone without using your hands by automatically sending when it detects audio input. The fact is that the VOX functionality is enabled in certain types of Baofeng UV-5R handheld units, which allows for the immediate detection of any kind of sound or speaking signal, leading to an automatic transmission even if the PTT button is not pressed. This feature helps those individuals who require talking on phones without being distracted when doing a task that needs their hands. The physical ptt button gets replaced by a VOX communication tool that enhances communication, user convenience, and efficiency.

These help cut down on interference and improve privacy in communication by only letting signals with matching tones or codes pass through the squelch and be heard.

REMOTE OPERATION AND PROGRAMMING

The Baofeng UV-5R can be manually customized using its buttons or automatically with software like CHIRP. Computer interfaces and remote control apps are two ways to do remote programming and operation.

Remote Function

1. Computer Interface: A programming cable, which is often a USB to 2-pin connector, can be used to connect the Baofeng UV-5R to a computer. Users can program the radio's frequencies, settings, and channels from their computer with the right software, making it simpler to manage sizable frequency and channel setups.

2. Remote Control Apps: Using a computer or smartphone, users can remotely operate the Baofeng UV-5R using a few third-party programs. These apps frequently establish a Bluetooth or USB connection with the radio, allowing users to modify channels, settings, and frequencies without having to physically touch the device.

Writing Code

1. Manual Programming: The menu system on the Baofeng UV-5R can be accessed to manually program the device. The radio's keypad and menu options allow users to directly enter frequencies,

tones, and other settings. While this approach works well for basic programming tasks, it can take a while for bigger frequency sets.

2. CHIRP Software: The Baofeng UV-5R is one of the many radios that can be programmed using this well-known, open-source application. To build, maintain, and edit radio configurations, users can download and install CHIRP on their computers. The software makes managing complicated radio sets easier by offering an intuitive interface for configuring channels, frequencies, and other parameters.

3. Internet Resources: A number of websites offer instructions and tutorials for configuring the Baofeng UV-5R. The support tools consist of video lessons, procedural manuals, and places where one can inquire about support from other members or share their experience. Online sources help users address complexities in setting up the device as well as troubleshooting.

BOOK 8: BAOFENG RADIO APPLICATIONS

HAM RADIO AND AMATEUR RADIO USAGE

Amateur radio, also known as "ham radio," is a non-commercial radio communication format that is used by licensed individuals for public service, leisure, or personal use. Government agencies have assigned different frequency bands to hams, which they utilize for local, national, and even international communication. This mode of communication is essential for radio communication growth, community outreach, experimentation, and emergency communication.

Amateur Radio with the Baofeng UV-5R

1. Frequency Bands:

The Baofeng UV-5R is appropriate for a variety of amateur radio activities because it covers both the VHF (Very High Frequency) and UHF (Ultra High Frequency) bands. The UV-5R allows hams to connect on many frequency bands, which allows them to take part in amateur radio contests, satellite communication, simplex communication, and local repeater networks.

2. Repeater Operations:

The UV-5R has access to repeater networks that expand amateur radio communications' reach and range. Ham radio operators can establish communication over extended distances and difficult terrain by setting repeater frequencies into their equipment. This capability is especially helpful in emergency scenarios where long-distance, dependable communication is essential.

3. Simplex Communication:

Direct point-to-point communication without the need for repeaters is possible with the UV-5R, which is available to hams. Local communication inside a defined area or between neighboring stations is a popular usage for this approach. The UV-5R is a good choice for simplex communication during outdoor activities, community events, or public service operations due to its small size and portability.

4. Emergency Communication:

In public service announcements and crisis situations, the Baofeng UV-5R can be a vital instrument for emergency communication. The gadget can be used by hams to create communication networks, transmit important information, and plan rescue operations in an emergency. Because of its sturdy qualities and small size, it is an invaluable tool for amateur radio operators involved in emergency response operations.

5. Programming Flexibility:

A vast array of frequencies, channels, and settings can be stored and managed by users thanks to the UV-5R's programming flexibility. Hams can easily arrange and update their radio settings with the aid of programming tools like CHIRP, providing smooth access to a wide range of frequencies and communication modes.

6. Community interaction:

Among amateur radio enthusiasts, the Baofeng UV-5R promotes community interaction. In order to promote a spirit of togetherness and cooperation among amateur radio operators, hams might utilize the gadget to take part in neighborhood clubs, amateur radio nets, and public service occasions. The UV-5R is a well-liked option for people wishing to participate in amateur radio activities and give back to their communities because of its accessibility and affordability.

COMMERCIAL AND PROFESSIONAL APPLICATIONS

Popular among amateur radio amateurs and professionals alike, the Baofeng UV-5R is a dual-band portable transceiver that is both versatile and reasonably priced. This tiny radio that started as an entertainment accessory is now used in numerous commercial and industrial applications due to its reliability, affordability, and wide range of features. Its versatility and user-friendly nature allow it to be used in numerous aspects as mentioned below.

1. Emergency Services and First Responders: The Baofeng UV-5R is a dependable tool for emergency services and first responders in circumstances when traditional communication routes may be unreliable or compromised. Its robust construction, small size, and long-range communication capabilities make it a valuable tool for disaster response efforts, search and rescue missions, and other urgent circumstances demanding prompt and efficient communication.

2. Event Management and Security Teams: To guarantee smooth communication throughout sizable crowds, concerts, festivals, and athletic events, event managers and security staff make use of the Baofeng UV-5R. Teams can work together more effectively because of its broad frequency range and configurable channels, which improve crowd control, emergency management, and general security.

3. Outdoor Activities and Adventure Sports: Hikers, campers, and outdoor lovers have their communication needs met with the Baofeng UV-5R. For wilderness excursions, mountaineering, and other outdoor sports when dependable communication is essential for safety and coordination, its sturdy build, extended battery life, and weather-resistant characteristics make it the perfect partner.

4. Construction and Industrial Sites: To promote effective communication among employees, managers, and supervisors on sizable project sites, construction firms, and industrial facilities use the Baofeng UV-5R. Its tough construction, hands-free functionality, and noise-canceling features allow for efficient communication even in demanding and noisy industrial settings, enhancing total output and safety procedures.

5. Transportation and Logistics: The Baofeng UV-5R facilitates continuous communication between drivers, dispatchers, and control centers in the transportation industry, which includes logistics organizations, trucking companies, and delivery services. Its dependable signal reception and long-range transmission aid in fleet movement management, delivery coordination, and handling unforeseen logistical issues, all of which improve operational efficiency and minimize delivery delays.

6. Hospitality and Retail Management: Employees in hotels, resorts, and retail establishments may maintain connections across floors and departments thanks to the Baofeng UV-5R, an effective communication tool in these fields. Discreet communication capabilities, such as the VOX function, and earbud compatibility, facilitate easy team cooperation, improving operational response and customer service.

7. Educational and Training Facilities: The Baofeng UV-5R is used by educational institutions and training centers to manage field trips, outdoor education courses, and hands-on training. Its user-friendly interface, extended battery life, and wide frequency coverage make it an invaluable tool for teachers and students, promoting efficient communication and guaranteeing safety during field trips and hands-on learning opportunities.

8. Community and Neighborhood Watch Groups: To improve communication and surveillance inside local communities, neighborhood patrols and community watch groups use the Baofeng UV-5R. Members may stay in continual communication thanks to its small size, extended battery life, and encryption features, which help to discourage criminal activity and foster a safer atmosphere for locals.

PUBLIC SAFETY AND EMERGENCY SERVICES

Although local laws and restrictions may apply, the Baofeng UV-5R has found use in a variety of public safety contexts. The following are some applications for this radio in public safety:

1. Search and Rescue (SAR): SAR teams frequently work in isolated locations with limited access to conventional communication. The UV-5R is a useful tool for organizing search and rescue operations because of its dual-band capability and configurable channels.

2. Community police: To improve communication between law enforcement and the community, certain community police efforts make use of the UV-5R. Due to its price, officers and volunteers can use it more widely.

3. Disaster Response: Communication infrastructure may be jeopardized during emergencies or natural catastrophes. First responders and disaster relief organizations can rely on the UV-5R as a dependable communication tool.

4. Event Security: To coordinate their efforts, security professionals frequently use two-way radios during large events like concerts, sporting events, and festivals. In these kinds of situations, the dual-band functionality and configurable channels of the UV-5R can be helpful.

5. Volunteer Fire Departments: The price of the UV-5R is advantageous for volunteer fire departments with tight resources. It gives firefighters an affordable way to communicate during training exercises and crises.

Suggestions for Public Safety Utilization

Although the Baofeng UV-5R has many benefits for applications in public safety, there are a few crucial things to remember:

1. Regulation and Licensing: The proper licenses may be needed in many countries in order to utilize two-way radios like the UV-5R for emergency services and public safety. Users need to be conscious of and follow local laws.

2. Interoperability: There's a chance that the UV-5R won't always work with the public safety agencies' current communication systems. Organizations should assess whether the UV-5R can interact with their current equipment as interoperability issues may develop.

3. Range Limitations: A number of variables, such as obstacles and topography, can affect the UV-5R's communication range. To increase the range in some circumstances, more repeaters or signal boosters can be required.

4. Sturdiness and Weather Resistance: Workers in emergency services and public safety frequently operate in harsh conditions. Making sure the UV-5R is shielded from physical harm, water, and dust is crucial.

5. Training: To optimize the UV-5R's efficacy in emergency response and public safety activities, proper training is a must. Users ought to be conversant with its characteristics and capabilities.

Apps for Emergency Services

Emergency medical services (EMS), fire, police, and other first responders are just a few of the many agencies and departments that fall under the umbrella of emergency services. With a few adjustments, the Baofeng UV-5R can be used in these environments:

1. Police Departments: The UV-5R's price and adaptability are advantageous to law enforcement organizations, particularly in the context of community policing initiatives. Officers can stay in touch with dispatch centers and other officers by using the radio.

2. Fire Departments: The UV-5R can be used by volunteer and small fire departments to improve firefighter communication in training exercises and emergency situations. Adequate instruction on radio protocols is essential.

3. Emergency Medical Services (EMS): To plan patient transfers, get in touch with hospitals, and guarantee staff safety, EMS teams can depend on the UV-5R. It offers an affordable means of communication.

4. Teams for Disaster Response: Coordinating quickly is essential in disaster-affected communities. Disaster response teams can communicate more effectively with local authorities and with each other by using the UV-5R.

5. Non-Governmental Organizations (NGOs): NGOs that work in areas with poor infrastructure are frequently engaged in humanitarian and disaster relief activities. In these situations, the UV-5R can be a dependable communication tool.

Personalization and Coding

Programmability is one of the Baofeng UV-5 R's main benefits. The radio can be configured by users with the precise frequencies and configurations required for their operations. Greater flexibility in the usage of the radio for emergency services and public safety is made possible by this adaptation.

1. Writing Software Code: Users can use the proper programming cable to connect the UV-5R to a computer and program it. There are numerous third-party software packages available to make programming easier. Users can adjust power levels, define channels, and change other parameters in this way.

2. Users should confirm that the frequencies they plan to use fall within the range that the radio supports. This is known as frequency compatibility. As different UV-5R models may offer marginally different frequency ranges, compatibility must be confirmed.

3. Use of Repeaters: Repeaters are a common tool used by emergency services and public safety to increase the radio's range. By programming the UV-5R to use repeater frequencies, communication coverage can be greatly increased.

4. Management of Interference: There may be a risk of radio interference in very populated regions. It is important for users to be aware of possible sources of interference and take precautions against any disruptions.

Battery Control

1. Careful battery management is necessary for the Baofeng UV-5R to be used in public safety and emergency services. Since the radio usually runs on a rechargeable battery pack, users should take the following into account:

2. Battery Life: Depending on the transmit power level and usage habits, the UV-5R's battery life may differ. Having extra batteries or a backup power source on hand is essential, especially for extended activities.

3. Battery Charging: Prior to every mission or event, users should make sure their batteries are well charged. To effectively charge several radios at once, several organizations make investments in multi-unit charging stations.

MILITARY AND GOVERNMENT USE

Baofeng radios, such as the UV-5R model, have been used sparingly in military and government applications, mostly in non-sensitive and non-critical settings. These radios are generally preferred due to their affordability, usability, and adaptability. It is crucial to remember that their usage in official military and government operations may be subject to stringent rules and restrictions, particularly in delicate situations where trustworthy and secure communication is essential.

Instruction and Non-Emergency Functions

The Baofeng radios can be used for non-tactical operations, non-critical communications, and training exercises in some military and government settings. They are a good choice for straightforward training situations and non-mission-critical coordination because of their price and configurable characteristics.

Use by Civilians and Auxiliaries

Baofeng radios can be used in conjunction with other communication technologies by some military and government branches for community involvement, disaster assistance, and outreach to the general public. In certain situations, their ease of use and programming capabilities may make them appropriate for non-mission-critical jobs.

Constraints & Points to Remember

It's important to be aware of the following restrictions and factors when thinking about using Baofeng radios in military and government applications:

Security and Encryption: In general, Baofeng radios lack sophisticated encryption features, which are important for government and military communications that need to be secure. It is essential to employ radios with strong security measures in sensitive operations to stop unwanted access to confidential data.

Interoperability difficulties Interoperability issues may arise when Baofeng radios are integrated into the military and government's current communication networks. complicated communication systems with the ability to function well in dynamic and complicated contexts are frequently needed for seamless coordination across several agencies and units.

Regulatory Compliance: Various government organizations oversee the usage of particular radio frequencies in a number of nations. In order to maintain legal compliance and prevent disruption of other vital communication routes, military and government organizations are required to abide by these laws.

Durability and Reliability: Government and military operations frequently take place in harsh conditions, necessitating the use of sturdy and long-lasting communication equipment. It's possible that Baofeng radios don't always adhere to the strict durability standards required for usage in challenging circumstances and high-stress settings.

Instruction and Acquaintance: For safe and efficient operation, Baofeng radios must be used with the appropriate training and familiarity with their features and limits. To operate the radios effectively and comprehend their place in the larger communication ecology, personnel should receive proper training.

RECREATIONAL AND OUTDOOR ACTIVITIES

The affordability, portability, and varied features of the Baofeng UV-5R make it a popular choice for recreational and outdoor users. It can improve safety, coordination, and enjoyment for a range of outdoor and recreational activities by acting as a dependable communication tool. The following are some typical outdoor and recreational uses for the Baofeng UV-5R:

1. Hiking and Backpacking: Hikers and backpackers can use the UV-5R to contact other members of their party or to signal for assistance in case of emergency when venturing into secluded or difficult terrain. It is the perfect partner for outdoor adventures because of its dual-band functionality and great communication range.

2. Camping and Caravanning: The UV-5R can help with communication between several campsites or cars during camping and caravanning excursions, which can improve coordination and guarantee everyone's safety.

3. Off-roading and trip Sports: Off-road fans can utilize the UV-5R to communicate with other riders regarding the route, possible hazards, or any crises that may occur throughout the trip. This includes ATV riders, dirt bikers, and 4x4 drivers.

4. Hunting and Fishing: To coordinate their efforts and share important information, such as game movements, fishing hotspots, or any unanticipated circumstances that need emergency assistance, hunters and fishermen frequently utilize the UV-5R.

5. Skiing and Snowboarding: Skiers and snowboarders may use the UV-5R to stay in touch with their friends while they're out on the slopes, which guarantees efficient coordination and communication—especially during inclement weather.

6. Outdoor Events & Festivals: To facilitate seamless communication among guests and help with event management, organizers and participants of outdoor events, festivals, and get-togethers can utilize the UV-5R.

Important Elements for Leisure Use

The Baofeng UV-5R is a versatile tool that is ideal for a range of outdoor and recreational activities because of its capabilities.

Frequency with two bands: Users can switch between multiple frequency bands with the UV-5R's dual-band capabilities to optimize communication depending on the particular terrain and surroundings.

Programmable Channels: Users can access several communication networks by using the radio's capacity to store many channels and frequencies. This feature facilitates connection with other outdoor enthusiasts and emergency services when necessary.

Long Battery Life: The UV-5R can have a longer usage period with appropriate battery management, guaranteeing constant communication during strenuous outdoor excursions or leisure activities.

Compact and Transportable Design: The UV-5R's lightweight and transportable design makes it easy to bring along on outdoor adventures without adding extra bulk to the user's equipment.

Emergency Features: The UV-5R's integrated flashlight and emergency alarm systems might come in handy when engaging in outdoor activities. They offer extra security in the event of unanticipated crises or low light.

The Best Ways to Use It Outside

The following best practices should be taken into consideration by users to guarantee the efficient and secure operation of the Baofeng UV-5R during outdoor and recreational activities:

Knowledge of Local Regulations: It is important to be aware of any local radio communication regulations or restrictions, such as allowable frequency bands and power output limits, before using the UV-5R outside.

Battery Backup and Charging: Carrying backup batteries or a portable power bank will assist in ensuring that the UV-5R remains functioning throughout the whole outdoor expedition, especially in remote places where power supplies may be limited.

Weatherproofing and Protection: By shielding the UV-5R from external factors like moisture, dust, and impact, protective cases or weatherproof pouches can help prolong its life and dependability while used outdoors.

Establishing unambiguous communication procedures, such as the use of call signals and standardized radio terminology, can help group members communicate more effectively and avoid miscommunication when engaging in outdoor activities.

Emergency Preparedness: Having a thorough emergency plan in place is essential before embarking on any outdoor trip. This involves teaching everyone involved the correct usage of the UV-5R's emergency features, as well as emergency protocols and meeting locations.

BOOK 9: BAOFENG RADIO LEGAL AND REGULATORY CONSIDERATIONS

LICENSING AND LEGAL REQUIREMENTS

Based on the area and region where the Baofeng UV-5R is being used, different licenses and laws may be needed to operate it. Following the rules and regulations set by the local regulatory authorities is very important to make sure you follow the law and avoid any legal problems. Here are some important things to think about when it comes to the Baofeng UV-5R's license and legal requirements:

Needs for a License

1. License for amateur radio:

People in a lot of countries who want to use the Baofeng UV-5R for amateur radio must first get the right license. To get this license, you usually have to pass a test that shows you know how to use a radio, follow the rules, and communicate properly.

2. Commercial License:

If you want to use a radio for business or commercial purposes, like by a security company or an event management company, you may need to get a certain license. This license makes sure that the business follows the rules set by the government and regulatory bodies.

Allocation of Frequencies and Rules

One important thing for users to know about frequency bands is that different types of radio contact use different frequency bands. The UV-5R should follow the rules set by the local telecommunications or regulatory body when it comes to frequency usage.

2. Limits on Transmission Power:

The transmission power that can be used in certain frequency bands may be limited in some countries. It is very important to stay within these power limits so that you don't mess up other radio conversations and so that you follow the law.

Limits set by the law

1. Things That Are Not Allowed:

Doing things that are not allowed, like sending illegal or offensive material, is strictly forbidden. Radio transmission laws must be followed by users, and they must also make sure that their use of the UV-5R doesn't break any laws or rules.

2. Mitigating Interference:

Users should take steps to make sure that authorized radio services and interactions are not harmed. This means not using bands that are meant for other specific uses and taking care of any interference problems right away.

Legal Obligations and Their Effects

1. Penalties for Not Following the Rules:

People who don't follow the rules and requirements for radio communication licenses may get fines, have their radio operation privileges taken away, or even be taken to court, based on how serious the violation was and the laws in the area.

2. Oversight by the Regulatory Authority:

The Regulatory Authority usually keeps an eye on radio communication operations to make sure that license and law requirements are met. Users should expect to be inspected and audited on a regular basis to make sure that their radio activities are legal.

FCC REGULATIONS (FOR THE UNITED STATES)

The regulatory agency in charge of monitoring and controlling all radio communication activities in the US is the Federal Communications Commission (FCC). In order to ensure effective spectrum management and reduce interference between various radio communication equipment, the FCC is essential. The Family Radio Service (FRS), General Mobile Radio Service (GMRS), and Amateur Radio Service are among the radio services that are subject to FCC regulations. These laws may have an impact on how the Baofeng UV-5R operates.

FCC Part 95: Individual Radio Services and FCC

Part 95 of the FCC regulations governs Personal Radio Services, which is what the Baofeng UV-5R falls under. The FRS and the GMRS, which are frequently used for leisure and personal use, are included in this category. Although an FCC license is not needed to access the FRS channels, using the UV-5R on GMRS frequencies requires a current GMRS license that must be obtained from the FCC. This license guarantees responsible operation within the allotted GMRS frequency bands and is necessary for regulatory compliance.

The Amateur Radio Service and FCC Part 97

An amateur radio operator may also use the Baofeng UV-5R in accordance with FCC Part 97 rules. Before operating the UV-5R within the approved amateur radio frequencies, radio operators must get the relevant FCC license, such as the Technician Class license. To guarantee the appropriate and legal use of the radio spectrum, the FCC establishes particular frequency allotments, power limitations, and operational guidelines for amateur radio operators.

Prohibition of Unauthorized Changes

It is noteworthy that the Baofeng UV-5R is not permitted to have any unauthorized changes made to it that would let it operate beyond its approved frequency ranges or power restrictions. This is absolutely prohibited by the FCC. Unauthorized changes could push the radio above the permissible limit and interfere negatively with other legitimate conversations. In order to comply with FCC restrictions, users should not make such modifications.

Compliance with Equipment Certification

The Baofeng UV-5R and all other radio frequency devices marketed and used in the US must abide by FCC equipment certification regulations. With this certification, you may be sure that the radio satisfies technical requirements and won't interact negatively with other permitted transmissions. For the UV-5R to operate legally in the US, compliance with FCC equipment certification requirements is necessary.

Mitigation of Interference and Reporting

It is anticipated that users of the Baofeng UV-5R will take preventative action to lessen any interference that may affect approved communications. When interference problems occur, consumers should notify the FCC right away so that it can look into them and find a solution. Users help to preserve the radio spectrum's integrity and guarantee a dependable, interference-free communication environment by actively reporting and mitigating interference.

Compliance with the Law and Penalties

Operating the Baofeng UV-5R legally and responsibly in the US requires adherence to FCC regulations. Serious consequences, such as fines, license suspension, or legal action, may result from breaking FCC laws. To prevent any legal repercussions, users must strictly adhere to the FCC regulations that apply to the UV-5R and ensure that they are completely aware of them.

INTERNATIONAL REGULATIONS AND LICENSING

It is essential to comprehend the varied regulatory environment that oversees radio communication in various nations and areas when utilizing the Baofeng radio abroad. Spectrum allocation, license requirements, international coordination, local law compliance, import and export rules, certification and compliance standards, and cross-border interference control are just a few of the many topics covered by these regulations.

Regulations and Spectrum Allocation

Various radio communication services are assigned specific frequency bands by each nation in accordance with its national regulatory structure. The spectrum allotted for the Baofeng radio's intended application in various nations should be known to users. It's critical to comprehend the acceptable frequency ranges in order to maintain legal compliance and prevent any unlawful use that might interfere with other communication systems.

License Conditions

Users of radio communication devices, such as the Baofeng radio, must get a valid license in several countries before using the equipment. The kind of service, the radio's intended usage, and the particular rules established by the national regulatory bodies can all influence the license needs. Users must obtain the required licenses to ensure legal compliance and be fully informed of the licensing requirements in the countries in which they intend to operate.

Observance of regional laws and ordinances

Users of the Baofeng radio must abide by all local laws and regulations governing radio transmission when operating the radio in different countries. These regulations may include a wide range of topics, such as limitations on power output, transmission protocols, frequency utilization, and material kinds that cannot be used. In order to ensure the responsible and legal operation of the Baofeng radio within the international setting, users should do a comprehensive investigation into and comprehension of the local laws and regulations.

International Arrangements and Coordination

The usage of radio communication equipment, such as the Baofeng radio, may be influenced by certain international agreements and coordination. It is important for users to be aware of any international agreements that can affect their capacity to use the radio in particular areas. In order to minimize cross-border interference and guarantee the effective and efficient use of the radio spectrum across several nations, international coordination is essential.

Regulations for Imports and Exports

Users need to know the import and export laws that apply to radio communication equipment in other countries while bringing the Baofeng radio across international borders. Following these rules is crucial to preventing problems at customs, the item being seized, or any legal ramifications that might result from breaking import and export laws.

Standards for Certification and Compliance

To make sure that radio communication equipment satisfies technical specifications and doesn't negatively interact with other communication systems, international standards organizations may create certification and compliance standards. Before using the Baofeng radio abroad, users should confirm that it conforms to all applicable international standards. Adherence to global certification and compliance guidelines is vital for the permissible and conscientious utilization of Baofeng radios in global settings.

Management of Cross-Border Interference

Users utilizing the Baofeng radio in close proximity to international boundaries should exercise caution as there exists a possibility of cross-border interference with the radio communication networks of neighboring countries. It is important to adhere to appropriate interference management procedures in order to reduce the possibility of interfering with authorized communications in nearby areas. Users must be aware of the consequences of cross-border interference and take proactive steps to reduce any possible interference problems that can occur when using the Baofeng radio in foreign environments.

FREQUENCY BAND USAGE GUIDELINES

INTERFERENCE AND COMPLIANCE

Considerations for Interference

When a signal is interfered with in radio communication, it means that it is being disrupted by other broadcasts, the environment, or technological constraints. Users of the Baofeng radio should be mindful of any interference sources and take the appropriate safety measures to reduce their influence. The following important variables increase the likelihood of interference:

1. Frequency Congestion: The likelihood of frequency congestion is higher in places with a large population density or plenty of radio communication devices. When choosing clear frequencies and

channels for their Baofeng radio, users should exercise caution to prevent interference with other approved communication systems.

2. Signal Overlap: When several radio devices are used in close proximity to one another, signal distortion and possible communication disruption might result. Frequencies and channels should be chosen carefully by users in order to reduce the possibility of signal overlap and avoid interfering with adjacent radio transmissions.

3. Environmental Factors: A number of environmental factors can impact the clarity and quality of radio signals, including air conditions, geographic obstructions, and sources of electromagnetic interference. When using the Baofeng radio, users should take certain ambient elements into account in order to lessen the influence of interference on the dependability of communication.

Adherence Mechanisms

Users of the Baofeng radio must abide by all applicable operational standards and regulatory requirements in order to guarantee its responsible and lawful functioning. Respecting compliance guidelines is essential to preserving effective communication procedures and lowering the possibility of illegal activity or fines. Several crucial steps for compliance consist of:

1. Guidelines for Regulations: It is imperative that users familiarize themselves with the regulations that control radio communication in their local areas. It is crucial for the legal and responsible operation of the Baofeng radio to comprehend frequency allocations, power output limitations, transmission protocols, and other regulatory requirements specified by local authorities.

2. Licensing Requirements: Operating the Baofeng radio may call for a specific license, depending on the intended usage and frequency ranges used. Before utilizing the equipment, users must get the necessary permission from the regulatory body to guarantee compliance with licensing requirements and prevent unauthorized operation.

3. Frequency control: To avoid interfering with other permitted communication systems, effective frequency control is essential. It is imperative for users to carefully consider and utilize the frequencies allotted to them, making sure that their radio broadcasts do not conflict with or interrupt other licensed radio services that share the same frequency ranges.

4. Adherence to Operational limitations: Users must make sure that the Baofeng radio complies with all operational restrictions listed in the regulatory standards and stays within the designated power output limitations. When these boundaries are crossed, there may be detrimental interference and a failure to adhere to regulations.

5. Designated Emergency Channels and Procedures: In order to guarantee that crucial communication networks continue to be available for emergency services and first responders, it is imperative that designated emergency channels and communication procedures be followed. In order to support the effective and dependable operation of emergency communication services, users should abstain from using emergency channels for non-critical communication.

BOOK 10: FUTURE OF BAOFENG RADIOS AND RESOURCES

ADVANCEMENTS IN RADIO TECHNOLOGY

Radio users favor the Baofeng UV-5R because of its affordable price, dual-band capability, and small size, which have all attracted a lot of attention. Being an old design, the UVR-UV5R has pushed the limit in portability through its many advanced features, which has caused it to achieve huge popularity among amateur and professional users. This has also assisted in enhancing radio communication through various means.

Dual-Band Capabilities and Adaptability

One key innovation offered by the Baofeng UV-R model is its dual-band capacity that allows users to select either VHF or UHF bands. The adjustable, broadband antenna makes the device compatible with a wide frequency range of operation, providing a convenient and smooth connection. Using various frequency bands has helped make communication easier and reliable to serve different types of users' requirements for both business and fun.

Small and Lightweight Design

Its compact and lightweight structure in the portable radio technology sets a new standard for comfort and easy usage. This is an ideal companion for all ham, public safety operators, and sports enthusiasts having to transport their equipment around. However, this gadget is mobile and it has an easy-to-use interface making it popular with many radiophonic users who are always on the move.

Increased Coverage of the Frequency Range

The Baofeng UV-5R has increased the potential for radio communication in a wide range of settings and applications thanks to its wider frequency range coverage. Users can connect to a variety of radio services, such as commercial radio frequencies, amateur radio bands, and public safety channels, thanks to its wide frequency range accessibility. Because of its wide frequency range coverage, the UV-5R is now much more versatile and useful, meeting the communication needs of a wide range of users in many sectors and industries.

Improved Amateur Radio Features

The UV-5R has been a great choice for many amateur hams using various frequencies due to its interoperability. Supporting licensed amateur radio operations has fostered an active and attentive user base among the amateur radio community. The UV-5R has integrated very well into the amateur radio network due to the improvement of the amateur radio's capabilities. This has produced an efficient and robust communication tool that users rely on for chats and ham radio services.

Economically and Availability

The Baofeng UV-5R has made a substantial impact on radio communication thanks to its accessibility and cost. Such low cost of the device has brought within reach even financially constrained individuals and institutions modern radio communication technologies. As compared to other technological breakthroughs, radio communication technology is relatively affordable hence it is easily accessible by people from various social backgrounds they benefit from its state-of-the-art attributes and features without having to compromise on its worth or standard.

A creative layout with an intuitive user interface

The UV-5R's layout and friendly user interface make operating and navigating its various features and functions a breeze.~ The simplicity in usage of this device is due to its user-friendly controls, clear display, and ergonomic arrangement of the controls and components. Hence, anyone can effectively get a hang of how to operate it without any prior knowledge. Radio users have readily adopted or received the UV-5R, mostly for its easily comprehensible interface which has improved radio interaction and happiness among them.

Conformity with Extra Attachments

The compatibility of the Baofeng UV-5R with a broad variety of extra accessories, including programming cords, battery packs, and external antennas, is another noteworthy innovation. The availability of such interoperability ensures that users can now expand and tailor the device's functionalities as they deem appropriate in order to meet their personal communication needs and idiosyncrasies. Additionally, it has also increased the UV-5R's capability to incorporate more accessories, ultimately allowing users to tailor their communication needs and meet varied field operations and situational demands.

Including Contemporary Radio Features

The Baofeng UV-5R is not the most technologically sophisticated gadget, but it does incorporate a number of contemporary radio technologies that have improved its overall performance and usefulness. Channel scanning, dual-watch functionality, emergency alarm systems, and programmable function keys are a few of these characteristics. The device's adaptability and usability have been improved by the incorporation of these contemporary radio characteristics, enabling users to participate in more streamlined and effective communication practices across a variety of operational scenarios.

Influence on Radio Communication Methodologies

Radio communication practices have been significantly impacted by the Baofeng UV-5R's contributions to portable radio technology, especially in the areas of amateur radio operations, outdoor leisure activities, and emergency response initiatives. Modern radio communication technology has never been more accessible or widely adopted thanks to its user-friendly design, reasonable price, and cutting-edge capabilities, which enable users to communicate seamlessly and dependably in a variety of social and professional contexts.

BAOFENG RADIO COMMUNITY AND RESOURCES

With a shared interest in Baofeng radio technology, users, amateurs, amateur radio operators, and professionals form the broad and interconnected Baofeng radio community. Members of this community may stay up to date on the most recent developments, industry best practices, and operational strategies pertaining to Baofeng radios. The community functions as a hub for cooperation, knowledge sharing, and skill development.

Internet Communities and Forums

The abundance of online forums and communities devoted to talking about Baofeng radio products, resolving technical problems, and exchanging experiences and insights is a pillar of the Baofeng radio community. Users can interact, ask questions, and share advice and best practices for making the most out of Baofeng radios on platforms like Reddit, social media groups, and specialist Baofeng user forums. Among Baofeng radio aficionados worldwide, these online forums promote a sense of camaraderie and mutual support through facilitating spirited discussions, knowledge sharing, and community interaction.

User Clubs and Groups

Baofeng radio clubs and user groups unite individuals who share a passion for technology and radio communication. To promote information exchange, skill development, and community building, these groups frequently plan events, training sessions, and gatherings both in person and virtually. In addition to learning from seasoned professionals, participating in practical activities pertaining to Baofeng radio operation and maintenance, and developing deep relationships with other radio enthusiasts, members of these user groups and clubs also help foster the expansion and vitality of the Baofeng radio community.

Tutorials & Resources for Education

A wealth of instructional materials, such as user manuals, tutorials, and instructional videos covering many facets of programming, debugging, and Baofeng radio operation, are available from the Baofeng radio community. With the help of these instructional materials, users should be able to make the most of their Baofeng radios and ensure dependable communication in a variety of operating scenarios. The Baofeng radio community encourages its members to learn new things on a constant basis by offering them thorough and easily available educational resources.

Online References and Records

When looking for in-depth knowledge on programming, setup, configuration, and advanced capabilities, Baofeng radio users can resort to extensive online manuals and documentation as useful resources. These manuals provide users with confidence in navigating the intricacies of Baofeng's radio operation by providing step-by-step instructions, practical insights, and troubleshooting suggestions. The Baofeng radio community supports self-guided learning and gives users the ability to make knowledgeable judgments about their radio communication practices and strategies by making trustworthy and organized documentation easily accessible.

Expert Guidance and Technical Assistance

Access to several technical support channels and knowledgeable guidance from seasoned specialists and experienced users is made possible by the Baofeng radio community. Users can get help fixing technical problems, setting up radio settings, and addressing performance difficulties through dedicated support forums, email assistance, and real-time chat platforms. The Baofeng radio community fosters a culture of cooperation and mutual aid through the provision of technical support and professional advice. This creates a welcoming atmosphere where users can seek advice and mentorship from competent peers and industry experts.

Networks and Associations for Amateur Radio

In the Baofeng radio community, amateur radio networks and associations are essential because they provide training courses, networking opportunities, and regulatory advice to amateur radio operators using Baofeng radios. These networks enable members to stay up to date on the newest advancements in amateur radio technology and regulatory compliance by facilitating the exchange of ideas, best practices, and industry insights. Amateur radio networks and groups promote active involvement, knowledge sharing, and adherence to established industry standards and procedures by creating a sense of community and teamwork.

Updates and Official Baofeng Documentation

Users of Baofeng radios can rely on official Baofeng material, such as device manuals, firmware upgrades, and software releases, as a trustworthy and authoritative source of information. Users may take advantage of new features and improvements for maximum radio performance by staying up to date on the latest product advancements, enhancements, and bug fixes thanks to access to official documentation and software updates. Baofeng promotes openness and user participation by giving timely and accurate product information. This builds a sense of confidence and dependability among the Baofeng radio community.

Regional and Worldwide Radio Events

The radio community has various occasions, local and global with regard to holding meetings and seminars whereby they can meet, interact with each other as well, and see what new things are developing in the world of radios. These events enable users to explore new products, technologies, and current trends in the market as well as network with the manufacturers, the industry's experts, and fellow radio hobbyists.

Involvement in regional and international radio events promotes professional growth, information exchange, and community engagement within the Baofeng radio community, strengthening the bonds of cooperation and solidarity among its members.

Amateur radio communities on Baofeng Radio Online

In the context of amateur radio, Baofeng radio online communities are crucial for knowledge sharing, skill development, and regulatory compliance among radio operators. These internet forums give users a place to talk about radio communication methods, exchange operational tales, and ask questions about industry standards and best practices. Baofeng radio online communities in amateur radio enable users to discover new vistas, increase their technical proficiency, and contribute to the development of amateur radio practices and technology by providing a collaborative and mentoring environment.

Social Media Networks and Interaction

Social media platforms function as vibrant centers where members of the Baofeng radio community may interact in real-time, exchange updates, and develop deep ties with other radio aficionados. Users can engage in debates, exchange ideas, and remain up to date on the most recent advancements in radio communication through social media groups, sites, and communities. Participation in social media within the Baofeng radio community fosters a feeling of inclusivity and community, motivating users to actively add to the pool of collective knowledge and take part in projects and activities that are driven by the community.

Workshops & Training Programs for Baofeng Radio

Users can improve their technical proficiency, gain a deeper comprehension of radio communication principles, and gain hands-on experience operating and maintaining Baofeng radios through training programs and seminars. Offering participants practical experience and immersive learning opportunities, these training programs and seminars cover a wide range of topics, such as radio setup, programming, troubleshooting, and emergency communication protocols. Baofeng radio training programs and seminars enable users to become proficient and confident in their radio communication capabilities and practices by promoting professional development and skill enhancement.

Contributions to the Open-Source Community of Baofeng Radio

The community surrounding Baofeng radios actively participates in the creation of open-source initiatives, software programs, and firmware upgrades that improve the features, compatibility, and performance of Baofeng radios. Expanding the capabilities and customization options accessible to Baofeng radio users is made possible by community-driven initiatives, cooperative development activities, and open-source contributions, all of which promote innovation and creativity within the community. The Baofeng radio community drives the advancement of Baofeng radio technology and practices by fostering a culture of shared learning, open communication, and continuous development through active involvement and cooperation.

Worldwide Baofeng Radio User Surveys and Response Systems

Global user surveys and feedback systems for Baofeng Radio offer insightful information on user preferences, satisfaction levels, and areas where the Baofeng Radio community may improve. Baofeng can make well-informed decisions on product development, feature enhancements, and customer support activities by gathering and evaluating user input. This allows Baofeng to better understand the requirements and expectations of its users and spot emerging trends and patterns. In the Baofeng radio community, user surveys and feedback systems are crucial instruments for encouraging user-centric innovation and advancing a culture of responsiveness and ongoing improvement.

Community Outreach Programs and Initiatives of Baofeng Radio

Programs and initiatives for Baofeng radio's community outreach are designed to raise awareness, educate, and encourage participation in radio communication among a variety of audiences, including the general public, educators, and students. These instructional sessions, open demonstrations, and community gatherings that emphasize the value of radio communication in public safety, disaster preparedness, and global connectivity are typical components of these outreach initiatives. Through encouraging a culture of radio communication literacy, empowerment, and social responsibility, Baofeng is able to cultivate a sense of community engagement and shared purpose among its partners and stakeholders. This is demonstrated by the outreach programs it has implemented in the community.

Community Code of Conduct and Ethical Standards for Baofeng Radio

Respect, inclusivity, professionalism, and a code of conduct are encouraged among the members of the Baofeng radio community. This code of conduct places a strong emphasis on the value of upholding a welcoming and encouraging community atmosphere, abstaining from discriminatory behavior, and abiding by legal requirements and industry standards. The Baofeng radio community works to establish a friendly and safe environment for all users, enthusiasts, and stakeholders by promoting a culture of mutual respect and accountability. This helps to build a sense of trust and camaraderie within the community.

Case Studies and Best Practices for the Baofeng Radio Community

The best practices and case studies of the Baofeng radio community offer insightful information gleaned from actual radio communication scenarios and applications. These resources provide useful advice and motivation for users looking to enhance their radio communication practices and tactics by showcasing effective use cases, tactical approaches, and creative approaches to Baofeng radio deployment and administration. The Baofeng radio community encourages a culture of constant learning, experimentation, and excellence by displaying excellent projects and success stories. This encourages the community to embrace industry best practices and standards.

USER GROUPS, FORUMS, AND ONLINE COMMUNITIES

A vibrant and interconnected network of enthusiasts, professionals, and hobbyists with a shared interest in Baofeng radio products, including models like the UV-5R, is formed by user groups, forums, and online communities for Baofeng radios. These platforms are excellent sources of information for community involvement, technical assistance, and knowledge exchange. They also create a cooperative atmosphere where users can share ideas, get guidance, and remain informed about new advancements in the Baofeng radio industry. A summary of various well-known discussion boards, user groups, and online communities catering to fans of Baofeng radios is provided below:

1. Reddit - Baofeng Radio Community: The Baofeng Radio Community is a well-known online community where users converse, exchange advice, and share technical and operational tidbits about Baofeng radios. In addition to providing a vibrant forum for user-generated material, community-driven projects, and exciting debates, this subreddit also helps Baofeng radio lovers worldwide connect and support one another.

2. Baofeng Radio User Forum: This lively online community provides a wide-ranging discussion board for Baofeng radio users to interact, work together, and share information. In order to meet the informational demands and preferences of a broad and international user base, this user forum offers a wide variety of subforums, topics, and threads covering different aspects of Baofeng radio operation, programming, troubleshooting, and best practices.

3. Baofeng Radio Enthusiasts Facebook Group: For fans of Baofeng radios, there are a number of Facebook groups that offer a forum for members to debate Baofeng radio technology, exchange stories, and pose queries. The growth and vitality of the Baofeng radio community on social media is facilitated by these Facebook groups, which encourage community engagement via connecting, networking, and collaboration on a range of radio communication projects and initiatives.

4. Online Communities Maintained by Amateur Radio Clubs: Baofeng radio users and amateur radio operators are served by the online communities and discussion forums that amateur radio clubs manage. Members can exchange ideas, take part in club activities, and keep up to date on the most recent advancements in amateur radio technology and practice through these online forums. A culture of cooperation, mentoring, and information exchange between radio enthusiasts and business experts is promoted by the incorporation of Baofeng radio talks into amateur radio club online forums.

5. Baofeng Radio Subsection on QRZ Forums: This section of the QRZ Forums provides a dedicated forum for users to discuss Baofeng radio products and related subjects, exchange ideas, and offer advice. For users looking for operational insights, troubleshooting assistance, and technical advice from seasoned professionals in the radio communication field, this area is an invaluable resource.

6. Baofeng Radio Groups on LinkedIn: These LinkedIn groups offer a professional networking environment where radio professionals, enthusiasts, and hobbyists may interact, exchange industry knowledge, and investigate joint venture prospects in the realm of radio communication. These LinkedIn groups encourage knowledge exchange, career growth, and industry networking, enabling members to stay updated on the latest trends, breakthroughs, and best practices in Baofeng radio technology and applications.

7. Baofeng Radio Discussion Boards on Eham.net: Eham.net has Baofeng radio discussion boards that cater to people seeking extensive information, user reviews, and technical insights on Baofeng radio products and accessories. These discussion boards serve as a vital archive of user-generated content and community-driven debates, giving a venue for users to express their experiences, recommendations, and issues relating to Baofeng's radio operation and performance.

8. Online Baofeng Radio Communities on Ham Radio Platforms: Ham radio platforms and communities provide dedicated sections and discussion groups focused on Baofeng radio technology

and applications. These online forums serve as hubs for users to engage in technical discussions, seek assistance from industry experts, and stay current on the latest news and developments in the world of Baofeng radios and amateur radio operations.

9. Baofeng Radio Forums on RadioReference.com: RadioReference.com has dedicated Baofeng radio forums that cater to people interested in public safety communications, scanning, and radio technology. By giving users a space to talk about Baofeng radio gear, programming methods, and other relevant subjects, these forums promote knowledge sharing and community involvement in the larger context of radio communication and public safety operations.

10. Baofeng Radio Groups on Discord: Discord hosts Baofeng radio groups and servers that enable real-time communication, chat, and collaboration tools for people interested in Baofeng radio technology and applications. These Discord groups offer interactive conversations, voice chats, and multimedia sharing, enabling members to connect, network, and collaborate on various Baofeng radio-related projects and activities, boosting community participation and shared learning among Discord users.

11. Baofeng Radio Online Communities on Twitter: These online communities on Twitter give people a place to debate Baofeng radio technology and applications, exchange updates, and have real-time interactions. These Twitter communities function as vibrant hubs where users can interact with prominent figures in the industry, take part in popular discussions, and remain up to date on news, events, and new product releases related to Baofeng radios and radio communication.

12. Baofeng Radio Subreddits on Particular Radio Applications: Baofeng radio technology and its role in supporting different communication needs and scenarios are frequently discussed and resources are linked to subreddits devoted to particular radio applications, such as public safety operations, outdoor activities, and emergency communication. These specialist subreddits provide users with tailored insights, use cases, and best practices for exploiting Baofeng radios in unique operational scenarios, creating a greater awareness of the diverse applications and capabilities of Baofeng radio devices.

13. Baofeng Radio Communities on Instagram: These communities provide a visual forum for individuals to exchange tales, pictures, and videos pertaining to the technology, applications, and use of Baofeng radios. These Instagram groups exhibit user-generated content, product demos, and behind-the-scenes insights into Baofeng radio operations, enabling community participation and visual storytelling within the framework of radio communication and technology.

14. Baofeng Radio Webinars and Virtual Events: These online forums offer consumers the chance to engage in real-time conversations, watch product demos, and listen to industry lectures about Baofeng radio technology and applications. Users can ask questions, interact with industry experts, and learn

firsthand about the newest trends, developments, and best practices in the world of Baofeng radios and radio communication during these online events.

15. Baofeng Radio Blogs and Online Publications: Baofeng radio blogs and online publications offer instructive articles, how-to instructions, and product evaluations that appeal to people seeking in-depth insights and analysis on Baofeng radio technology and its varied uses. These internet resources are invaluable sources of industry knowledge, professional viewpoints, and firsthand accounts from users, giving readers a thorough grasp of how Baofeng radios are changing and how that is affecting radio communication as a whole.

RECOMMENDED READING AND ADDITIONAL REFERENCES

For anyone seeking greater information and in-depth knowledge on Baofeng radios, including the UV-5R model, a number of recommended reading materials and other references are available to provide thorough insights into the world of Baofeng radio technology. These resources cover many elements of Baofeng radios, including operation, programming, troubleshooting, and best practices, allowing users a broader grasp of the functioning, applications, and possibilities of Baofeng radio devices. Here is a curated collection of recommended reading and extra references for amateurs and professionals interested in Baofeng radios:

1. "The Baofeng UV-5R Beginner's Guide" by Andrew B.

This thorough guide serves as a crucial resource for persons new to Baofeng radios, offering readers a complete introduction to the fundamental principles of radio communication, Baofeng UV-5R capabilities, and operating best practices. The book provides step-by-step instructions, practical examples, and troubleshooting tips to help beginners navigate the basics of Baofeng radio operation, programming, and maintenance, making it an indispensable guide for users seeking to establish a strong foundation in radio communication practices and principles.

2. "Baofeng UV-5R Ham Radio" by Robert C.

This informative handbook concentrates on the applications of Baofeng UV-5R in amateur radio operations, giving users thorough insights into the device's capabilities, programming functionalities, and antenna optimization techniques. The book covers complex topics like frequency management, signal propagation, and radio communication protocols, allowing readers a deeper knowledge of the role of Baofeng UV-5R in supporting amateur radio activities and fostering communication within the amateur radio community.

3. "Mastering Baofeng UV-5R Programming" by Jennifer S.

This programming-focused reference is meant to help users grasp the intricacies of Baofeng UV-5R programming, offering readers a full introduction to programming techniques, software tools, and customization options for enhancing radio performance and usefulness. The book digs into advanced programming concepts, channel setups, and firmware updates, offering readers practical insights and expert guidance on exploiting the full potential of the Baofeng UV-5R for a wide range of communication applications and operating scenarios.

4. "Baofeng UV-5R for Emergency Communication" by Michael D.

This specialist guide focuses on the role of Baofeng UV-5R in emergency communication scenarios, offering users practical ideas, case studies, and real-world examples of using the device for building dependable communication networks during emergencies and crisis situations. The book covers topics such as emergency channel programming, distress signal protocols, and communication network management, providing readers with crucial knowledge and best practices for deploying Baofeng UV-5R as an effective communication tool in critical and life-saving situations.

5. Jessica M. Baofeng's "Baofeng UV-5R Radio Communication Handbook"

This comprehensive manual serves as a complete introduction to radio communication principles, techniques, and best practices, with a specific focus on Baofeng UV-5R operations and applications. In order to give readers a comprehensive grasp of the underlying ideas and real-world factors that guide successful radio communication techniques, the book examines the principles of radio signal propagation, antenna design, and communication network architecture. The manual provides readers with a comprehensive understanding of the complex dynamics of Baofeng UV-5R radio transmission by fusing theoretical understanding with real-world examples.

6. David R.'s "Baofeng UV-5R Advanced Troubleshooting Guide"

This advanced troubleshooting guide is intended for users who want to improve their technical knowledge and ability to solve problems with regard to the operation and maintenance of Baofeng UV-5R radios. The book presents in-depth analysis, diagnostic procedures, and expert insights into resolving complicated technical issues, signal interference concerns, and performance obstacles that users may confront in diverse operating scenarios. Through providing useful answers and troubleshooting techniques, the guide gives readers the skills and information they need to properly handle important radio communication problems and guarantee continuous operating performance.

7. Baofeng Technologies' "Baofeng UV-5R User Manual and Technical Guide" (Official Documentation)

This official user manual and technical guide, issued by Baofeng Technologies, provides a complete reference for users seeking extensive information on the Baofeng UV-5 R's features, functions, and operational specifications. The guide provides users with in-depth information on radio setup, configuration options, and programming methods, delivering authoritative insights and dependable guidance on enhancing the device's performance and compatibility with various communication systems and protocols. Users may make sure they have access to correct and current information about Baofeng UV-5R operation, maintenance, and regulatory compliance by consulting the official user manual and technical guide. This will help them make well-informed decisions and get the most out of their radio communication experience.

8. Baofeng Radio Online Documentation and FAQs (Online Resource)

Through the internet website that offers online documentation and FAQs for Baofeng radio models such as UV-5R, one can get a lot of information on different technical matters and the solutions for possible problems using the said devices. If one is looking for answers on how to get through some common operational problems associated with using the Baofeng radios, exploring the online documentation and FAQs available on the Baofeng Technologies official website would help one unlock a goldmine of resources such as user guide manuals.

CLOSING REMARKS AND FUTURE OUTLOOK

As we come to the end of "Baofeng Radio Bible - 10 Books In 1: The Ultimate Guerrilla's Guide," we think back on the useful information and deep understanding that was shared in every chapter. This handbook will be relevant not only to neophytes but also to professionals intending to derive maximum benefits from Baofeng technology. It goes over everything from the background and development of Baofeng radios to advanced features and legal issues.

Radio technology is always getting better, and there are more and more ways to use and apply Baofeng radios. The future looks bright for these radios. Amateur radio operators as well as the individuals working in public safety and emergency services often find it practical to use Baofeng radios. They are dependable and can undertake numerous functions. Baofeng radios are set to play a key role in promoting connectivity, improving communication efficiency, and ensuring safety and security in a wide range of operating settings as the world of radio communication continues to change.

The active Baofeng radio community and its huge collection of user groups, boards, and online communities are also great places to share information, work together, and keep learning. With the active use of these resources and thorough knowledge found here, people can track the latest trends, practices, and technology that concern Baofeng radio. By doing so, they will understand radio communication principles and their use in practice.

We encourage people to read the suggested readings and other sources mentioned in this guide because they provide useful information and expert opinions on different areas of Baofeng radio technology, use, and upkeep. By staying aware and taking an active role in how they use Baofeng radios, users can improve their communication experiences, deal with difficult issues, and help radio communication practices and technologies continue to grow and improve.

With respect to the future of Baofeng Radios, it is a world of endless opportunities and opportunities that can change your life, where advanced qualities are merged with enhanced operations and a friendly interface. Armed with such knowledge about the concepts and methodologies discussed herein, users embark on an exploration, fresh ideas, and continual progress towards reshaping the tomorrow of Baofeng radios and how they tie people up, interact, as well as engage each other in myriad

CONCLUSION

We've embarked on a journey through the evolution, features, operation, maintenance, and legal aspects of Baofeng radios. As we reflect on the wealth of information gathered across these ten books, it becomes evident that this "bible" serves as an indispensable resource for both beginners and seasoned users of Baofeng radios.

BOOK 1 laid the foundation, introducing us to the intriguing history and diverse models of Baofeng radios. Understanding why these radios are a good idea and identifying the target audience set the stage for the in-depth exploration that followed.

BOOK 2 delved into the intricate features and functions of Baofeng radios, unraveling the mysteries of radio frequencies, channels, and codes. The intricacies of the display and keypad, along with audio and volume controls, were expertly decoded.

BOOK 3 provided a hands-on guide to programming channels and memories, utilizing scanning and dual watch features, and mastering emergency and NOAA weather alerts. This book equipped readers with the knowledge to navigate the diverse functionalities of their Baofeng radios seamlessly.

BOOK 4 expanded our horizons by exploring a range of accessories, from headsets and microphones to carrying cases and holsters. The chapter on upgrading and modifying Baofeng radios empowered users to customize their devices according to their needs.

BOOK 5 elevated the discourse, emphasizing radio etiquette, communication protocols, and effective techniques. It tackled privacy, security concerns, and adeptly addressed troubleshooting common communication issues.

BOOK 6 shifted focus to maintenance and care, offering insights into battery management, charging, storing, and protecting Baofeng radios. The importance of firmware updates and software management for optimal performance was underscored.

BOOK 7 delved into advanced features, unraveling the potential of cross-band repeating, voice-activated operation, and remote programming. These advanced functionalities opened up new possibilities for users seeking to maximize their Baofeng radio experience.

BOOK 8 explored the diverse applications of Baofeng radios across commercial, professional, public safety, military, government, and recreational domains. The versatility of these radios became evident, showcasing their relevance in a myriad of scenarios.

BOOK 9 navigated the complex landscape of legal and regulatory considerations, elucidating FCC regulations (for the United States), international regulations, licensing, frequency band usage guidelines, and the crucial aspects of interference and compliance.

Finally, BOOK 10 offered a glimpse into the future of Baofeng radios and valuable resources for users to stay connected with the ever-evolving community. The recommended reading and additional references, coupled with closing remarks and a future outlook, serve as a testament to the dynamic nature of this technology.

I hope that this has been a comprehensive guide, a beacon illuminating the intricate world of Baofeng radios. May this collection continue to serve as a valuable companion on your journey in exploring, understanding, and mastering the art of Baofeng radio communication. Happy radio adventures!

BONUS

BONUS 1

BONUS 2

BONUS 3

BONUS 4

BONUS 5

Made in United States
Troutdale, OR
02/10/2024